The Magic of Watches

A Smart Introduction to Fine Watchmaking

To Yannick and Philippe

The Magic of Watches

A Smart Introduction to Fine Watchmaking

By Louis Nardin

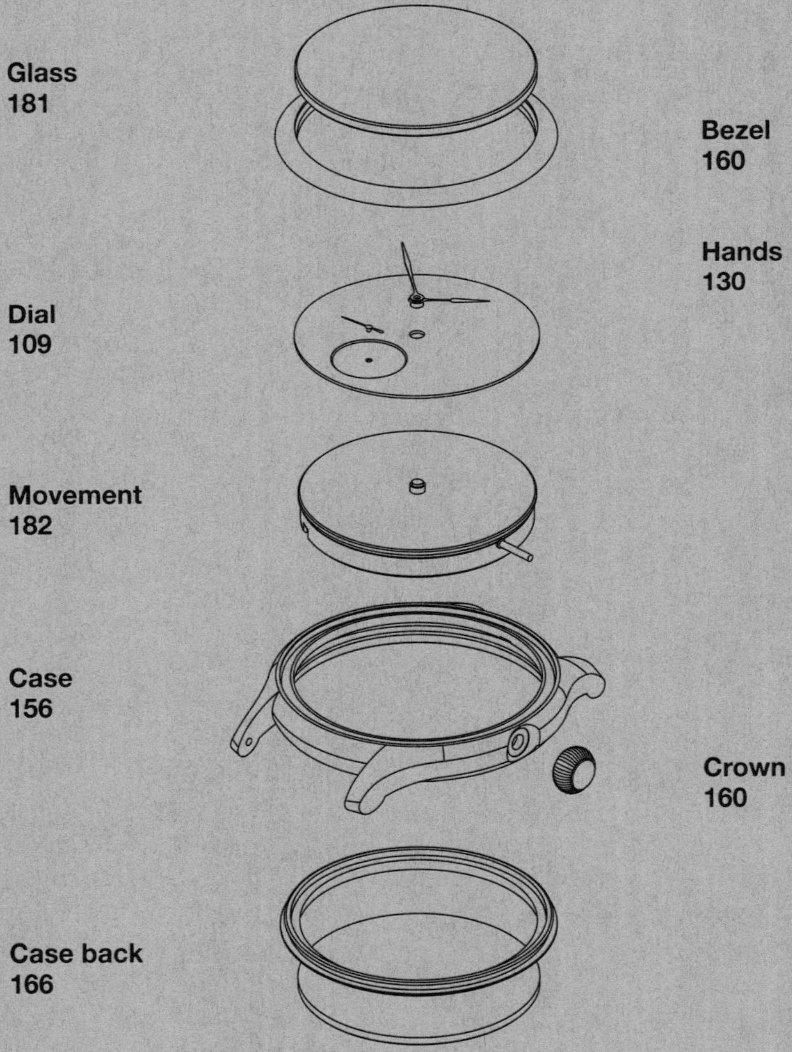
Glass
181
Bezel
160
Hands
130
Dial
109
Movement
182
Case
156
Crown
160
Case back
166

Strap
168

Lugs
164

Aperture
127

Pusher
163

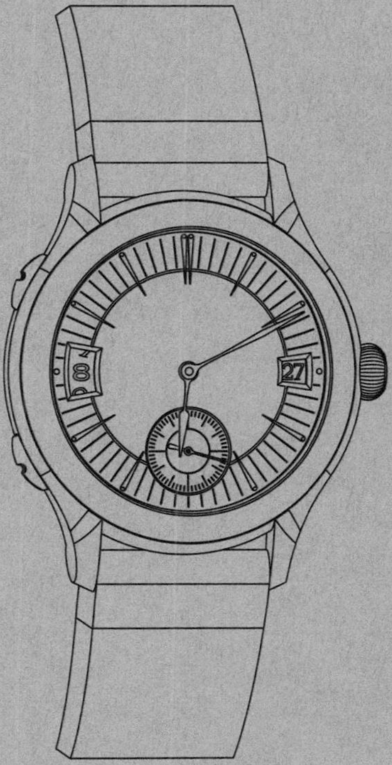

Indexes
114

Minute scale
115

Table of contents

Welcome
09 — 29

Buying
30 — 71

Care
72 — 80

Anatomy
105 — 203

Welcome

Dear Reader,

Do watches intrigue you? Are you attracted to them but still in the dark about their inner workings and the secrets they hold?

Thank you for joining this discussion to get to know watches better and to come to understand why watches are magic.

A watch is more than just a timekeeper. It belongs to a universe of luxury as huge and rich as the realms of wine, cars, art, and fashion—full of variety, creativity, performance, bravura and delight.

In addition, watches can subtly express "who you really are."

So let's begin our journey—it's going to be a great one.

12

The Magic of Watches

Why a book, why this one

This book is the result of what happened to me in 2012 and 2013. The reasons it exists are closely linked to a long journey I undertook at this time. But my personal story with watchmaking is far older than that. I'd like to tell it to you just briefly.

Rather an exception in the cosmopolitan city of Geneva, where I live now with my family, I'm a pure Swiss guy coming from one of the major centers of the watchmaking region, the city of La Chaux-de-Fonds. It is located in the Jura mountain range that extends from Geneva up to Schaffhausen. Nestled within its hills and valleys are almost all the Swiss factories and workshops dedicated to watchmaking. But you can't guess what is going on inside these factories as an outsider looking at kilometers of austere walls in rather deserted streets. To know the place, you must get introduced and be invited within. If you are lucky enough to have this opportunity, you are just on the verge of one of the most incredible experiences of your life. This is where specialists can transform rough metals and other materials into tiny and splendid mechanisms holding a soul within. This is also where you'll find rows of million-dollar machines designed to tool thousands of parts and also complete timepieces, like the colorful Swatch watches, for instance. And, if you are really lucky, it's also where you may steal a glance at ultracomplications and highly valuable timepieces with golden cases, paved with diamonds, and perhaps adorned with artisanal realizations like enamel dials. Even as a native of the region, I knew nothing about such things growing up—I only recognized the names of the brands appearing on the headquarters' exteriors.

To be frank, I never got interested in these little ticking items before I emerged from my studies in history and sociology and went out looking for a job as a journalist. Fate led me into this world, as if by chance—the way every good thing should happen in life. I got introduced to the glorious realm of watches by a highly respected journalist, Michel Jeannot. He hired me and taught me the ins and outs of writing for the watchmaking field. I owe him a lot. My knowledge and interest grew as I became editor in chief of Worldtempus.com, a website specializing in watches and watchmaking news. After three years, I left for a ten-month trip; I needed to find my own image of the world.

The idea for this book clicked during that long trip. While traveling and meeting folks from everywhere, I naturally got asked very often about my job. And what I remember perfectly as I answered is the faces of these people with huge question marks in their eyes. "Is it possible to write only about watches?" and "What is so interesting about watches that you can earn your living only writing about them?" were common questions. This was my first clue that it would be worthwhile to pen an introduction to this unique world. The second push came from most of my friends in Switzerland—friends who ignored everything concerning these little marvels. When I talked about my work, I could often see in their eyes a sparkle of interest, but just as often I saw the boredom that set in when they got lost in my description of the latest amazing piece. This was also a great motivator. Then I began to also consider offering tips about how to choose and buy a watch. And that is how and why I decided to write this book as a manual of sorts. To create it, I applied my own knowledge, but I also interviewed specialists in sales, design, and publishing. And several experts also worked directly on the manuscript—and I can never thank them enough for their contributions.

Thus, *The Magic of Watches* must be considered as an introductory essay that can naturally be improved upon by delving further into watchmaking. Its unique ambition is to be a useful toolbox for the newcomer and to give him or her the keys to this amazing universe. Indeed, everyone deserves to build an individual opinion regarding these beauties.

Talking brands names and sponsors

The purpose of this book is not to advocate for any particular brand, but to give the amateur a basic education to come up with an individual opinion. Therefore, "The Magic of Watches" is not linked to backing by any brand. The funds for its production have come only from institutions or individuals convinced of the usefulness of the project. No deals have been made with any brand, cited or not. Consequently, the brands noted here are mentioned because they represent a significant, reliable, and fully approved example for connoisseurs on the topic.

Why try to keep to the strict minimum in citing brands in a book about watches? Isn't that like talking about the best wines without mentioning Château Lafitte or Romanée-Conti? Brands are like fixed stars; they help us find our way in the galaxy of possibilities when we buy things. So why should we skip them and focus on theory rather than providing examples?

It's because all brands are not equal. You have the historical, the new, some that are incredibly famous, and others that deserve more of the spotlight than they receive, and so on. It is impossible to find the right balance within their constellations. Moreover, commercial competition is fierce among them. Giving an advantage to one or the other wouldn't be fair.

Why watches are magic

Watches are magic because they are precious, beautiful, technical, and still useful—to list just a few reasons. Yes, in an age in which the time is constantly visible on our cell phones, computers, and other screens, you shouldn't need a bracelet to tell you what time it is. Paradoxically, your watch is still your best companion because it provides only this information. This is the first point: a watch is simple. It doesn't matter how many functions it has, they will always be limited compared to a miniaturized computer. You don't have to scan a screen to spot the time nor explore the countless applications of your phone to access it. No, with a watch, the transmittal of information is direct, easy, and natural. And that's not even taking into account the frantic feeling when you don't remember where you put your damned cell phone! Yes indeed, it's so much easier to just glance at your left or right wrist to read the time without connection or electricity. This is the reason, pragmatically speaking, why watches are still necessary and useful tools today. You might also be surprised by how much you'll miss it when you don't wear one.

Talisman

Watches can make us dream and feel good. Yes! This little object is much more than an accumulation of its parts. It's also an expression of ourselves—of our tastes, of our personality. A watch tells a lot about the person who wears it. It's also a fashion item. It links people together—creating informal contemporary tribes. It can be an instrument of personal achievement because it might be precious, if not very precious. Why not translate a professional or personal success into something valuable that you can keep close to you, anywhere and anytime? Yes, your watch is one of your talismans, an object that in a way helps to record your life. Beside your wedding ring, if you have one, this is the one object with which you will spend most of your time.

Technology and personality

Watches are also magic because they are complex, hard to build, and sometimes hard to find too. Did you know that the simplest mechanical movement contains no fewer than 150 parts? And every one of them must be produced with a precision of one-thousandth of a millimeter? Did you realize that you can find, in some extreme models, materials otherwise only used in spaceships and Formula 1 cars? And what about the fact that if you change the battery when it's low, a standard plastic watch might still last dozens of years without any other care, until the plastic itself begins to disintegrate?

From the most basic to the most complicated, proper watches combine the utmost in technology, unique and strong design, and also carry the personality of a brand. Just as every vineyard specializes in a particular kind of grape grown under certain circumstances and providing a specific taste to a wine, every brand has its own, often strong personality to create the necessary aura around the product. So, considering this, let's get acquainted with the subtleties and particularities of what truly makes a good watch.

How to use this book

This book is not a novel, but rather a guide or a manual. It can be read from beginning to end, but also perused freely by picking out topics that you are interested in, in the order you prefer. So, use it as you will because there's no specific way to enter into the culture of watches.

The Magic of Watches is divided into five independent sections.

1
Welcome

The introductory section will teach you briefly about the history of timekeeping and what a wristwatch is, from the origins of the field until today.

2
Buying

What use is knowing about watches without having some clues as to how to acquire them and which one to choose? The second section provides tips on just such questions and includes some general concepts that each potential buyer should bear in mind. It also shares an overview and a presentation of the main kinds of watches. The secondhand market and how to sell a watch—to free up funds to buy the next one!—will also be considered.

3
Care

A watch needs to be cared for, and that is the focus of the third section. You will learn how to look after your watch, how to fix it—or get it fixed—if necessary, and some good practices to make it last. You will also find out some tips for traveling and how to insure your watch.

4
Anatomy

The fourth section is a comprehensive glossary explaining the various elements of a watch. It describes how they work as well as the different kinds of movements and functions.

As you will also notice, a good watch can come from many places around the world, and unknown brands can offer excellent products. So, keep your mind open and consider the facts rather than be seduced by small talk or ad campaigns. What you want is really the best value for your money—and you can have it.

5
Conclusion

The indispensable farewell comes next. Remember above all to have a lot of fun in your exploration of watch culture!

The first mission: To tell the time

Tracking of time began with the Neolithic era around 9000 BCE in Asia Minor when humankind stopped living by only hunting and gathering and started to grow crops and domesticate animals. In other words: when agriculture developed and wandering stopped, time became significant. The importance of predicting the seasons became crucial to ensure harvests and survival. The solution was found in the sky: the first human beings noticed the repetition of astronomical cycles such as the path of the sun and the alternation of day and night.

Historic testaments like dolmens, isolated menhirs, and basic circular monuments are visible traces of the primitive approach to timekeeping since 5000 BCE. Our ancestors soon also noticed celestial objects crossing the night sky on a regular basis and at various frequencies, starting with the moon. Astronomy, then, can easily be said to be the mother of time measurement. This origin must never be forgotten, even when speaking about contemporary wristwatches.

Once it caught humankind's attention, savants never stopped inventing and building tools to improve the measurement of time. Why is that? In a nutshell, "controlling" time allowed ways to improve survival as well as structure society. For instance, in 1541 Protestant reformist John Calvin established punctuality as a fundamental social value—thereby ensuring that people would be on time for religious celebrations, as well as work, a fundamental value for this religious current.

The challenge of measuring time

Through the centuries following the advent of agriculture, investigations and experiments have led to the creation of smaller and handier instruments able also to display time ever more precisely. The water clock, or clepsydra, has used water to indicate periods of time since at least 2000 BCE. Sundials display the hours using the shadow of their gnomon—usually a stick fixed on a dial. From the incense timer and the first wooden movements all the way to the atomic clock, a lot of options have been explored.

As previously stated, the measurement of time is intimately connected with the observation of the stars and their motion. And humans tried rapidly to invent machines able to reproduce their movement. Indeed, creating a smaller system showing how the planets move is the necessary precursor to a machine able to reproduce the passing of time in a structured way. The first of these was the Antikythera mechanism found at the beginning of the twentieth century in a shipwreck close to the Greek island of the same name. Created in about 200 BCE, it was the first geared mechanism able to reproduce celestial motion and to predict astrological positions as they relate to an existing calendar.

Hundreds of years later, during the fourteenth century in Padua, Italy, scientist and clockmaker Giovanni Dondi invented an animated planetarium which combined a clock movement with the indication of the position of the sun, the moon, the motion of five planets, and the tracking of religious celebrations. This instrument, a fundamental marker in the quest for accurate timekeeping, was important for the quantity of information it provided. It also indicates the high-water mark of the savants of the time as well as the state of mechanical timekeeping.

As religion has historically been such a strong social force, it comes as no surprise that the first real clock mechanisms were installed in churches at the end of the Middle Ages. One of the most famous of these is the astronomical clock in Prague. Installed in 1410, it displays twenty-one different indications and is the oldest one of its kind still active.

In parallel, watchmakers worked on miniaturizing movements to make the clocks wearable, thereby transforming them into watches. These were close to being jewels because of their adornments, which involved several crafts coming from the jewelry sector. From the fifteenth to the eighteenth century, watchmaking experienced a scientific boom with the discovery of all the major inventions that are present in watches still today. Several watchmakers—like Abraham-Louis Breguet, John Arnold, and Ferdinand Berthoud—left their mark on this time mainly by improving the precision of movements.

Watches were becoming more accurate, more reliable, and integrated more functions. Their production also increased significantly as demand exploded because of the growth of commercial exchange as well as new means of transportation such as the train during the second half of the nineteenth century.

Even though the first wristwatches were already appearing in the seventeenth century, they only became really popular at the beginning of the twentieth. They were first used on battlefields as officers needed to know the exact time in order to synchronize their actions. Shortly thereafter, civilian society also began to take note of them. Jewelers helped a great deal in transforming timepieces into creative and attractive wearable objects by inventing new shapes and finding ways to decorate them. In short, watches transformed into real luxury objects that would reflect the status of their owners. A single gesture—extracting the wrist from one's sleeve—and those around you would immediately know who you really are.

At the beginning of the twentieth century, watches were still mechanical, and great improvements were underway to make them stronger, more resistant, and more reliable. Consequently, wristwatches eclipsed pocket watches and demand increased for them. In Switzerland, the switch occurred in 1930 when wristwatch production outstripped that of pocket watches.

The extras: Complications

As soon as they could, watchmakers extended the functionality of clocks and then watches with additional elements called *complications*. In addition to displaying time, these timekeepers could also, for instance, indicate the phases of the moon or, later, periods of time counted on demand with a chronograph function.

A complication is defined as any function that goes beyond the standard indication of time in hours, minutes, and perhaps seconds. Directly powered by the movement, these common and popular supplementary functions can be very numerous. Throughout time, the challenge has been to miniaturize them until they could fit in a wristwatch. But the love for the challenge that complications represent by themselves also led to the creation of superlative machines. And this quest, which must be taken essentially as a purist's passion for watchmaking perfection, is still going on. For instance, the Vacheron Constantin pocket watch Reference 57260 launched in 2015 is the most complicated portable timepiece at the moment with no fewer than fifty-seven complications, all mechanical of course.

1960s

The quartz revolution

In the 1950s, engineers managed to introduce electricity in a wristwatch. Electricity, considered a type of constant force no matter the conditions, is the ideal power source for a watch. The earliest concepts started to appear in the nineteenth century, but the first person to patent an electric clock was Alexander Bain in 1841. His concept used electromagnetic pulses to animate and control the movements of the pendulum, which is the regulating component providing a clock with its precision thanks to its regular oscillations. But a really interesting discovery for watchmaking took place in 1928 when the first clock utilizing a quartz oscillator was invented by Warren A. Marrison. Quartz has the physical property of perfectly regulating an electrical flux. Integrated into a watch, it helps provide a much more uniform measurement of time.

During the second half of the twentieth century, the goal was to bring electrical control to wristwatches. In Switzerland, engineer Max Hetzel invented the first electronic oscillator designed to fit into a wristwatch movement for the brand Bulova. The watch, called the Accutron, launched in 1960.

Meanwhile, some Asian countries, and Japan in particular, saw their economy booming because of the ability to generate strong technical innovations. Soon brands and watch corporations launched the first wristwatches with quartz oscillators. The gold medal went to Seiko, a Japanese watch brand that became a giant industrial group better known today as Seiko Epson. It presented the Astron in 1967, the first watch to integrate a tuning-fork quartz crystal oscillator connected to a stepping motor.

In Switzerland, engineers were working hard to stay in the race as Asian products broke all the records in terms of precision and pricing. A consortium called CEH (Centre Electronique Horloger) presented for instance the Beta B21 quartz movement in 1969, so called because it involved twenty-one partners. The first Swiss quartz movement produced industrially, it would end up being utilized only in a few models, with some brands launching collections around them.

In reality, the Swiss watchmaking ecosystem paid too much attention to pure scientific research. There was no strong and common plan prepared to fight the assault of the numerous, well-priced, extremely accurate Japanese quartz timepieces. Customers chiefly needed and therefore wanted the most accurate watches, which could only be achieved through the use of quartz movements.

Recent historical research says that the Japanese production system was custom designed for mass production. Simultaneously, the Swiss production system was suffering from a high level of cartelization that prevented the industry from being reactive and efficient. Seiko was in fact able to propose the most accurate watches in large quantities and for a reasonable price thanks to a strategy based on volume. On the other side of the globe, the Swiss industry torturously looked for an appropriate response.

1980s

Fantastic plastic

The rebirth of the Swiss watch industry occurred in two phases. First, production was finally rationalized. Many firms, however, lost their independence and were integrated into a structure that could be profitable in the modern era. This move gave birth in particular to what is today known as the Swatch Group, the strongest watch-making consortium in the country from an industrial point of view. The Swatch watch embodied this historical change in Switzerland's production tradition, and experts agree that this legendary timepiece helped to preserve the industry's facilities. It also made watches accessible due to the use of plastic and low production costs. But there is more. Thanks in particular to the Swatch watch era, wristwatches became colorful, trendy, provocative, and fun as artists were called upon to express their creativity by designing original models. The low price point made it possible for consumers to acquire more than one Swatch as a fashion accessory. And this is also what started a very strong fan base for collecting watches.

1990s
From an instrument to much more

The second half of the twentieth century ushered in the digitalization of the world. Computers and electronic items became instruments of everyday life. This has radically modified the relationship people have with their watches. As the time became available on every screen—on computers, cell phones, digital panels in the streets, microwaves, ovens, and other appliances in the home—watches should have become obsolete. But instead a shift occurred. The primary role of a watch continues to be an instrument to accurately tell the time, but what accounts for more interest now is its role as an artifact—a precious object that combines price, status, and style with a strong interest in tradition and technology.

Like never before, watches have become objects reflecting the owner's personality, taste, and wealth. They have also become wonderful platforms for creativity. In addition to being luxurious, they are now totally fashionable items. This metamorphosis, linked with huge improvements in development and production processes and the use of new, innovative materials, has led to a revolution in watchmaking. In addition, since the 2000s, a general interest for authentic and purely mechanical objects has matured. Marketing departments now call the shots for each brand, with a mission to make dreams come true and excite customers in a totally new way.

This (r)evolution has led to the resurrection of many historical brands. Among them, we can cite Bovet and Arnold & Son as interesting examples. They built their comebacks by calling upon strong and distinctive elements of their respective pasts: from richly decorated and complicated pocket watches through objects designed for the marine and military fields, all the while keeping track of precise timekeeping. Many new brands have also popped up, taking advantage of improved machining tools, new powerful conceptualization programs and strategies that open up a wide range of possibilities. This approach has taken on particular significance in the high-end and luxury segment. Greubel Forsey, De Bethune, and Richard Mille are some relevant names in terms of this type of creativity.

In one century, the wristwatch has undergone a fantastic transformation, most obviously within the luxury segment. But don't be put off: every price range underwent its own metamorphosis. There is no doubt that the perfect wrist-watch exists for everyone in this day and age. The most important step you can take in finding yours is to unleash your curiosity and open up your desire to learn the secrets of a good watch. Then all you have to do is apply your new knowledge when the time comes to make your choice.

2010s
"Smart," or connected, watches

The Internet has changed all our lives. There is no need to quote names, digital companies are everywhere and their presence has totally modified interactions between people as they create valuable new goods: among them databases, of course, but also devices and services. The strongest players are those who have developed the smartest and most efficient environments with devices allowing easy online access. The new gadgets are great: useful, easy to use, offering so many possibilities. So, why do watches, and in particular mechanical ones, still exist? As we have discussed, they offer something different, something more—much more—than surface value. The "Ten Reasons Why" section of this book coming up next exists to analyze these values.

Watchmaking companies are wholly aware of the necessity to be part of the modern era and bring long and relevant experience in terms of wearable devices to the table. The challenge is to create or transform watches in order to make them compatible with digital environments. Such new, connected watches are very different from traditional watches in one main point: duration. A traditional watch is made to withstand the test of time. But even lower-quality timepieces stick around long enough so that you will have to at least change the battery several times or take them for repairs before moving on to a new one. This is a major difference in comparison with modern connected devices, which lose their relevance as soon as they become too weak to run the latest applications and functionalities—a significantly shorter time span. Consequently, we don't develop the same kind of attachment toward them that we do for mechanical watches.

Buying

You must be feeling impatient by now. This section will help you in the process of acquiring your own watch. You now understand what roles a watch plays today. Here is where we get into advice and tips for making a good choice whether you are buying a new watch or delving into the secondary market. We will also go deeper into the motivations that transform a watch lover into a collector. In the end, the goal is a qualitative overview of why billions are spent each year on buying wristwatches.

Ten reasons why you should get a watch

In general, people don't know how interesting, appealing, or even necessary a watch can be. That is normal. The world is full of beautiful things, and it is just impossible to know about all of them. But if you have decided to read this book, it is because you have some interest in watches.

So just go through the points below and focus on the ones that resonate for you. They might be the gateway to understanding your interest.

1

An art form

According to the Oxford Dictionary, art is: "the expression or application of human creative skill and imagination, typically in a visual form such as painting or sculpture, producing works to be appreciated primarily for their beauty or emotional power; (or): Subjects of study primarily concerned with human creativity and social life."

There it is: watches fit this definition. They speak more to feelings and emotions than to rationality. A watch is first and foremost an object that inspires, opens the doors to one's creativity, acts as the incarnation of desires—becoming a pilot, an astronaut, an explorer, a movie star—that have been concentrated into something precious.

Actually, this evolution in meaning is recent. We will go through this in detail later, but a single watch may be at the same time an ostentatious symbol, a statement, a testimony of one's education, a piece of history, and more. It is too narrow to consider a watch as only an instrument or a gift. In fact, the definition of a watch includes many subjective and personal factors. Fashion, trends, and cultural behaviors have also modified its status. Emotions have become a key factor. That's certainly why, today, professionals are keener to consider watches a "twelfth art."

2

Little treasure

The fact is that humans always seem to need some "extra" pieces of ecstasy. Watches offer that kind of fun. In addition, there is always something novel and fascinating going on with the new models, materials, brands, innovations, incredible stories, records, and celebrities.

This notion of pleasure combined with preciousness makes the watch equivalent to a jewel. Since the prehistoric era, humans have created and worn artifacts of many functions such as religious symbols, social markers, and rare and desirable objects. Nothing has changed except, after thousands of years of evolution, humans have acquired the ability to create smaller and more refined objects.

As jewelry, the watch fits especially well for men concerned with their looks and who like to distinguish themselves with a proper style—hello, neodandies! Even a simple model is a testimony of personal taste with a particular value. So if a super basic watch is already something special, what would a more complex and valuable model represent? And if you add precious materials such as gold and gems or craftsmanship like engraving or miniature painting and an extremely high level of technical finesse, could the result be anything less than a jewel? In addition, there is the emotional link that forms when a buyer gets his or her watch that makes this comparison even stronger.

At the moment someone buys a watch of his or her own, this excited feeling wells up. The watch becomes precious not only because of its value; it becomes important because, for most, it embodies the duration between the onset of desire and the act of buying. Sometimes, the retailer doesn't have the chosen model in stock, or none among those presented pleases the buyer enough. Patience then starts to sound like a mantra. If the desired model must be ordered and there is a strong demand for it, it can be weeks, months, or even years before the consumer can obtain it—and I'm not only talking about unique pieces made on demand.

The importance of a watch can also be tied to the circumstances under which it was received. Graduations, weddings, and all kinds of successes are special moments when a watch can go from a gift box to a wrist. Looking at the watch will forever remind its owner of this particular moment.

Speaking literally about the passing of time, a watch is also a testimony of every moment of one's life. Indeed, a watch counts every second of our existence. It's the buddy that accompanies you faithfully, always, everywhere. Symbolically, it records your life. In that light, if you decide one day to bequeath your watch to your child, your offspring will inherit a part of your personal history.

3

Authenticity

We now have access to apparel that can assume plenty of functions, including displaying time. First came the smartphone, which allows its wearer to manage tons of tasks and organize life. With it, the world became more accessible, practical, and easier.

These machines that can do everything are basically portable computers. When they are switched off or the batteries die, they are dead. Then our access to the world is abruptly cut off, and we have to survive on our own. This feeling of an absolute void is not easy to deal with and generates the need to take back control of one's life. It's not that the new technology must be rejected with a trip back to the past. The issue seems subtler: the world is accelerating, making the past disappear even faster. But humans need history, roots, and elements that reassure them. Old items inspire confidence and evoke memories. This kind of sweet nostalgia is what the vintage trend—now in evidence for a whole decade—exploits. It's becoming the main tendency—and the watch industry is certainly not immune. Consequently, one object indicating the time with little chance of failing in its mission is a real ally. It can even evoke some inspiring times. This kind of object may be more than welcome on our wrists!

4
Value

Is getting a watch really worth it? Or, in other words, shouldn't I put my money into something else? The value of any object is defined by many factors: objective ones like the cost of production or the materials used and less pragmatic ones such as the prestige of the brand or the rarity of the product—something that, in watchmaking, can create a strong interest and competition among fans and make prices explode.

As in other industries, watch companies set the prices of their products considering all these elements.Today, the positioning of the brand—the price segment in which it sells its watches—is one of the most important factors. Nevertheless reputation, history, market presence, and visibility on significant watch platforms such as mass media and events can be considered valuable clues about the objective worth of a watch.

On a different note, buying a watch as if one were making an investment is playing a risky game. Take the secondary market for vehicles as an example: almost any watch loses at least one-quarter of its value the moment it leaves the shop.

At the same time, it is true that some watches sold at auction reach stratospheric prices. Just think of the more than twenty-four million dollars paid for the Patek Philippe "Graves" pocket watch in 2014! But these are exceptional circumstances created by rarity, special specifications, and highly positive momentum on the market.

The right attitude is to expect to pay a certain price for a product that is of a quality that inspires you. It's almost impossible to expect more from fine watches than the great happiness and pride you feel when you wear the object of your dreams. If you discount that as a valid reason, just remember how many people buy watches every day. A product attracting hundreds of millions of customers per year is not a peripheral phenomenon. In the end, the real value of a watch is located in the eyes and the heart of its owner. If you love it, then it will be worth the cost.

5

Beauty

Defining beauty is an art in itself. It could be the pleasant feeling one experiences while appreciating an object or a representation that releases harmony, excellence, truthfulness, and/or originality, as the Oxford Dictionary states. This entry asserts above all that beauty remains firmly in the eyes of the observer. Thus, beauty is not something natural or commonly shared; it is a mix of culture, education, and personal taste. Consequently, beauty is as subjective and different as the number of humans in this world. In watches there are several elements that combine to create beauty, and each of them must be taken into account. We can obviously point out the design that will seduce at first glance—or later the shapes and the functions.

But there is more to it than that. Did you know, for instance, that the name Ebel comes from a love story? The artificial word is the combination of the names of the brand's founders: Eugène Blum and his wife, Alice Levy. So, wearing a beautiful watch is first and foremost embellishing yourself. If it also evokes important moments, then it becomes a sensitive object to which we can't stay indifferent.

6

Personality

Who are you? What do you want to show the world about yourself? In a discreet way, a watch reveals facets of its owner's character. You are what you wear. So what are you going to wear?

As with any item you are wearing, you will feel more comfortable if there is a balance between your watch and yourself. Some aficionados are even luckier if they own several models and can pick the one that reflects their mood of the day.

In fact, finding the watch that reflects your personality takes time, patience, and willpower. Friends and professionals can be helpful in this quest. Don't give up until you're fully satisfied and happy with your choice. One day, people will pay attention to what is on your wrist.

7
Passion

Watches can initiate intense passion because of their design, complexity, rarity, age, value, and as artistic objects. Over time, the multitude of models, knowledge, and exchanges between aficionados have created a culture surrounding watches above and beyond what could be termed "an exclusive club." Recently, thanks to new tools of online communication, the phenomenon has become more intense globally. Some forums dedicated to watches as well as specialized media have existed for a couple of decades now. So, getting an informal education on watches is not a problem and helps in understanding better the refinements of a timepiece—until simple leisure becomes a true mission.

8
Patrimony

The heritage dimension of a watch matters a lot. Wearing the watch of one's grandfather or grandmother makes us feel closer to him or her. It is also a piece of family history that goes from one generation to the next. Thus the motto Patek Philippe, the most prestigious traditional independent brand, has used since 1996 is no coincidence: "You never actually own a Patek Philippe. You merely look after it for the next generation." So, if you are looking for something particular you might leave for posterity, put a watch on your list.

9
Dream machines

A watch can make you dream. It works like a little window to a minuscule universe in motion. Some people love observing the details of the design, the mechanism at work, the wheels in motion, and the pulsations of the hairspring on a mechanical watch. For those who are keen on architecture, science fiction, industrial realization, or art, it can be a fascinating and even hypnotic show. The metronomic pulsations of the movement can also have a relaxing effect. It's the regular motion of the movement's parts that catches the attention and encourages your mind to roam.

10
Perfect gift

Watches make perfect gifts, and many are sold every week for this purpose. What is nice is that there are watches for all tastes and all budgets. Some people like to give inexpensive, fun watches, most often made out of plastic, as spontaneous presents. When the price goes up, it's often a gift for a milestone or a special event: religious celebrations, reaching voting age, graduations, weddings, professional citations and awards.

But giving a watch as a present is also risky. Please be sure of your choice or be ready to have the watch returned to the shop. It is a better solution to bring the recipient to the retailer than to try to make the surprise of his or her life.

Buying
Ten reasons
Choosing
The two fundamental questions
Feelings and thoughts
Price matters
Brands
Women only
Customization
Key elements
Secondhand
Selling your watch
Knowledge

Choosing

We have just been through the reasons—and the desires—that can lead you to buy a watch. Now you have to make your choice. And this has more to do with a slow maturation process than simply crossing a threshold and picking out the one you like more.

To be certain to hit the bull's-eye, details matter, and each of them must be considered with due attention. Hands, dial, lug shape, colors, choice of quartz or mechanical movement, kind of strap, comfort, look, how you will use the watch—they all add up.

"Everything to me and about me" or the two fundamental questions

Before becoming too pragmatic and chasing the right specs and the best deal, you should first understand why you want to buy a watch. You might challenge yourself regarding the reasons and emotions that are pushing you to this purchase. Such soul-searching could be helpful for making the right choice and avoiding potential regrets!

There is no bad or good reason here, just something strong enough to guide you. Of course, you can buy a watch spontaneously. It is accomplished in a heartbeat, and you might be happy with it for years—most watches are sold that way. But if you are not easily moved by emotion, then enjoy taking your time until you find "the one." In both cases, two questions define your relationship with the watch: What does the watch say about me? And what does the watch say to me? You will save time—and certainly money—if you find the appropriate answers in advance.

To me

You will spend a lot of time with your watch on your wrist, and you will look at it many times over the course of a day. Thus, this watch really has to inspire you. What does it say to you,? How do you feel when you look at it? How does its beauty give you energy, joy, and pride? These questions have the purpose of getting you to create a list of a few useful keywords. These will be important criteria while looking for the right watch and work as guiding stars during moments of doubt.

About me

Your watch is an expression of yourself, of your personality. It reflects you, what you like, provides clues about your way of living and your fashion sense. Through your watch you can decide which part of your personality you want to share with others. It speaks for you. Colorful, technical, classical, exotic, or rare—it gathers intimate and precise features of your personality and puts them on your wrist for all to see.

Feelings and thoughts

It is easy to fall in love with a watch and be almost obsessed with it—and unsurprisingly brands have perfectly learned how to master the art of seduction. Crucial, as we all know, is that each beautiful love story starts with a deep heartbeat. But passion and feelings can also unfortunately lead toward some painful and expensive mistakes.

That is why a watch should be chosen in two stages. First, you listen to your emotions regarding the model and become aware of the desire it provokes—or not. Then, it is time to do your homework by gathering information about it until you can make a rational analysis. This two-pronged approach will allow you to experience the deep happiness of putting your beloved watch on day after day.

A love story

Each love story begins with fireworks; if the romance continues, it moves on to become a warm fire. This analogy also works for buying a watch. Here we share a few ideas related to the emotional and irrational aspects of buying a timepiece that should help make your watch love story as intense as it is beautiful.

The quickening heartbeat

You look at the watch of your dreams and reason just flies out the window. It is love at first sight. The sparkle and racing pulse are necessary ingredients in choosing a watch you adore. The essence and strength of this wonderful feeling are a key indicator. Once you've felt this pull, at least you know how special and strong this "watch feeling" can be. It becomes a very helpful sign for the rest the process. A watch will have to generate at least this level of emotion to get on the short list.

The look

Of course, a watch must first of all please you visually. Professionals estimate that 60 percent of sales and perhaps even more are made strictly based on design. This means that all other elements—brand, functions, comfort, etc.—are secondary. As for the quickening heartbeat, take note of it, but don't stop there. A watch is more than a cute look.

Dealing with design is complicated. If you have no experience with watch design, you can start developing your eye by surfing the web or checking out specialized magazines.
Choose your preferences among the models you see and expand this group by adding new ones. Continue refining your selection by adding/deleting models until you are satisfied. You have just created a panel displaying your own particular taste in watches. Now, step back and observe each watch longer to discover what pleases you so much. Maybe you'll notice that you are first interested in one kind of metal, a type of shape for the hands, or certain combinations of functions. If you find that you like certain details, make a list of them; they will be really helpful!

Reputation of the brand

A watch is always one reflection of a brand. So get introduced to the brands, explore their worlds, pick out what makes them special for you, and take your time to find the ones you really like. Choosing a watch from a brand that you esteem will strengthen its value.

Chats and stories

Be curious and ask around about the watch you are targeting. Friends and acquaintances might have been through this before and can help. Chatting informally about watches is good because there is no marked path to take to enter this little universe. Getting others' impressions, advice, and opinions is beneficial. Have a look online and pick inspiring watch websites to get a feel for how important watch decisions can be for passionate people. Hours can be spent discussing the design of a dial or the history of a function, mechanism, or indication. For a beginner who is keen on digital tools, a watch forum is a good place to ask precise questions and obtain information. So get some feedback and follow the eventual advice only if it speaks to you. Don't be shy and speak frankly. It is not stupid to ask about basic things, and it is not provocative to say that you don't appreciate something.

Being rational (analysis)

The very technical nature of a watch opens up a large field for analysis. Materials, function, and performances—there is a lot to investigate. After exploring subjective and subtle notions, let's move to an objective and pragmatic reality.

Styles and uses

As there are no official standards on the categorization of watch styles, the following ones were conceived to be simple and clear. They aim to show the variety of styles and to offer up a short description in order to highlight what makes each of them special.

One more thing: Although watches are (still) mainly male jewelry, things are changing. The various elements discussed in the following sections are valid regardless of gender.

— Elegance and classicism

If you were asked to draw a watch, how would it look? It would probably be a circle with two different-sized sticks meeting in the middle. This can be considered the basic symbol of a watch: time expressed in the simplest and purest way. This design remains the simplest and the most sold. Why? Because it embodies the essence of a timepiece. From its tremendous simplicity comes its elegance. As with a perfectly tailored suit, every element fits ideally and nothing is superfluous. This concept is fundamental, and many artists share this point: perfect realization is reached when nothing more can be removed. For this reason, designers spend hours trying to reshape this basic traditional and classic architecture. And the result is hours of work and tons of sketches just to change a few details. Classical design is a very specific approach.

A state of mind focused on purity and refinement can define models usually stamped "classical." These are resolutely anti-bling watches dedicated to one single mission: indicating time in the purest and most comfortable way.

Such a model should actually belong to the beginner as well as to the connoisseur. Indeed, there are some days when you don't want to put anything on your wrist, but a classical watch will always be acceptable. And after having tried various other kinds of watches, a moment will always arrive when you want to come back to basics.

— Sports and extreme conditions

Exploration and competition have boosted watchmaking like nothing else. Many famously robust and technical models have sprung from these two fields. As an example, accurate and solid watches have become essential for racing, skiing, sprinting, diving, and climbing. Because of the extreme conditions and specific uses involved, special functionalities and designs have been invented. Some have become iconic.

Of course, these watches are for those who like competition and pushing the limits. Sometimes their design is more colorful and their size very often bigger than that of classical watches. And for a few years now, more and more women have found they like to wear these muscular pieces.

Today, the basic quality standards in watch manufacturing make almost any watch as strong as a sports watches (apart from really specific models). However, brands continue to have a healthy appetite for innovation and remain on the lookout for any improvement that would leave a watch stronger and more efficient.

— Complicated

Complications are additional functions beyond the basic hours, minutes, and seconds. To sum it up, two main categories can be established among complications: the simple and the very complex. Let's focus here on the complex ones as they are much more difficult to produce than the others. And for this reason, they generate much more attention—until they almost totally monopolize the concept of "complicated."

Watchmaking aficionados love high complications for many reasons. They are technically fascinating, rare, and express a brand's level of craftsmanship as well as its mechanical philosophy. They also evoke a certain historical context, sometimes taking a particular role in the fascinating odyssey of timekeeping that involves cultural, scientific, and technical aspects in particular. In addition to engineering, the movements containing these complications are most often highly decorated.

As you may have guessed by now, highly complicated watches are also much more expensive. As they represent the pinnacle of luxury in watchmaking, their inaccessibility adds to their prestige. But don't get discouraged. Sometimes the secondhand market and auctions offer good surprises.

— Artistic and precious

Some watches are jewels or pieces of art that also display the time. Nothing is too good for them, starting with materials. Precious metals such as gold and platinum share space with gemstones like diamonds and sapphires. The artistic touch comes from the talent of the craftsmen as well as from special decorative techniques used to create these watches. The result must be as unique as it is evocative.

Some of these watches represent a fusion between horology and jewelry. The bracelet might be a river of precious stones with the watch being just one element belonging to a jeweled set comprising earrings and a necklace. Other models are more art oriented. The dial often serves as a basis for delicate pictures executed using several artisanal techniques. Like highly complicated watches, the very best artistic watches remain almost unreachable for the majority of us. They are very often unique pieces or belong to very small series. The only chance to have a look at them is to go into boutiques. Don't be shy: these realizations are so outstanding that it is really worth pushing through the door to have a look at them. More affordable versions usually exist too, which are made using other techniques, often involving machines as well as fewer precious materials—which doesn't make them less pretty and desirable.

— Contemporary

Like every big family, watchmaking also has its crazy cousins. Since the middle of the 1990s, young and talented watchmakers and creators have started to veer off the traditional path to realize what they conceive of as new watchmaking. Thanks to their audacity and improvements in conception and fabrication tools, they have invented completely unexpected timepieces that look like spaceships or work like kinetic sculptures. Selling their watches at the same sort of prices as very complicated and exclusive ones by traditional brands, they have attracted a new generation and also moderated the rigidity of tradition. Now well-established, they continue to surprise.

"Contemporary watches" also refer to timepieces with modern designs created by artists or designers that are not specialized in the field. For them, a watch is an industrial object like any other and the general rules of design can apply. This approach is interesting—but most of the time not satisfying. The connection between form and function in a watch is so particular and specialized that creative people with experience are most often really more able to deal with it.

— **Vintage**
Back to the roots! As in fashion, watch trends explore the past in order to invent the future. Surfing the vintage wave, for a while now brands have been regularly rediscovering their attics and cellars. Actually, the trend is strong because older watches have become more and more collectable; this market has grown significantly over these last few years. During auction sessions you can witness epic bidding wars between collectors as the best lots reach skyrocketing prices—sometimes as much as several million Swiss francs for a single wristwatch. Many connoisseurs, however, attend auctions to obtain much more affordable models. In any case, all brands have started to reinterpret their iconic historical models. Some have mastered this process so well that they have even become famous for it. Vintage-inspired watches are often great choices as they offer classic looks outfitted with modern quality standards.

Adequate material
Your watch's case can be made out of different materials. Basically, the choice falls between metal and nonmetal, precious and not precious. Of course, metals are stronger than plastics, and sometimes lighter—like titanium. Luxurious models are naturally made of gold, platinum, palladium, and even silver. Metal watches can also be repolished to become shiny again and erase the scratches. They remain the first choice.

Nonmetallic watches are interesting because most of them are, by default, made out of different kinds of plastics. In particular, this allows a great range of creativity in terms of color and design.

Ceramic has a kind of special status. It is, for instance, a good choice for sports and design models, as it is strong, scratchproof, and better resistant to shocks. Historically speaking, the earliest fabrication process is intriguing; a zirconium dioxide powder is heated and then put under great pressure in a mold to create the part. In the early days, that meant that only certain shapes were possible, and creative designers had to play with such technical challenges. Rado has built its entire identity on this material for instance. Today, almost every shape is possible in ceramic.

Carbon fiber is also frequently employed for its light weight, strength, and look. Its fabrication process involves weaving fibers cast in resin, then compressed together at a high temperature.

The current norms have become strict: all materials used for cases, straps, and so on must be hypoallergenic to protect the owner's skin.

Price matters

How much should I pay? And why is it so expensive? These are the million-dollar questions to be answered! A good electronic watch costs about twenty dollars and works perfectly. On the opposite end, an ultracomplicated mechanical timepiece will be sold for hundreds of thousands, and its accuracy won't be as good as that of the quartz watch. Why the price difference then?

Many aspects come together to impact the price of a watch, and precision is not the most important by far: the wristwatch is no longer intimately related to its initial mission. It has become a fabulous object bearing many different senses and meanings. The question of price should be treated from a different perspective by asking yourself first the question of how much you are ready to pay for a watch. Because the good news is that today you can find good models for every kind of budget.

Brands

What makes a good watch brand, and why should I trust it? As with the price, this issue is not black or white. Brands based in countries with strong watchmaking traditions like Switzerland, Japan, and Germany take advantage of a dense industrial network, of strong know-how, and of public and private infrastructures specially dedicated to this industry. All of this creates a positive environment that helps increase the quality. But product quality is not the whole picture. The quality of service is also crucial, meaning the time you may have to wait if the watch is not in stock or for after-sales service matters, for example. In this case, big brands, or brands belonging to major luxury groups, have good structures to manage production and services. This power is also positive because it works as insurance: the brand will not disappear easily.

When considering life cycles of companies, those of a historical nature are particularly interesting. Even if many changes may have occurred, the continuity has a positive influence. Nevertheless, do pay attention: an old name can also hide a brand that has been dead for many years that has been relaunched. In such a case, you have to regard it as a recent brand.

A fresh, new brand always brings something different to the mix. As success is not guaranteed because competition is hard in this industry, be cautious. To clear yourself of any doubt, the best strategy is to dig deeper. In other words, inform yourself enough about the solidity of the project, its presence worldwide, the duration and costs of maintenance and service, and check on some online forums, for example, to see what other customers are sharing about the brand and their experiences with the products.

Women only

Most women consider watches differently than men do. For them, the instrument status and symbol of power that males generally focus on are unimportant. They are keener to appreciate the watch as a precious accessory. Earrings, necklaces, and rings are a watch's friend in the minds of many women. Consequently, the look is essential, whereas the technology might be less so. In fact, women prefer watches to be practical. That is why quartz models are still very much appreciated in this market. As long as the battery is charged, it will not stop ticking and tocking because you forgot to wear it. That is the point: many women say they have no time for winding and setting a watch. And for some of them, the crown primarily represents a risk of damaging fingernails. Most women like to be able to take a watch out of their jewelry boxes and wear it whenever they want.

But other women enjoy horological technology more. In this case, their attention often goes to watches with complications that are not purely linked to the measurement of time. Instead of a chronograph or a perpetual calendar, they might prefer a moon phase or another aesthetic and imaginative complication.

Customization

Customization is no stranger to watchmaking. From changing the strap to entirely transforming a watch, everything is (theoretically) possible. Nowadays, aficionados are keen on being able to change bracelets or bands themselves as these are relatively easy to remove and replace and remain affordable to the majority. Custom straps and bracelets are essentially made out of leather or textile (like a NATO strap, for example). But not all watches can accommodate "unofficial" straps. So before you make your order, check the size and length to see if your attachment system allows it.

Some firms have become specialists in deep modifications of watches. These interpretations can boast special colors, additional indications, and even funky dials. It is a special and quite intimate market that essentially concerns iconic models from famous brands. But be aware that brands automatically void a watch's guarantee if such modifications are not made by them. And the fact is that they obviously don't sell these kinds of transformed pieces. That's why the most affordable personal customization remains engravings on the back of a watch, which make it special with a short message that always stays close to its owner.

Key elements

At the threshold of choosing a watch, some points are particularly important to consider. The following are the best to take into account to find "The One." Some of these key elements concern the watch itself, others the act of buying the watch.

Aesthetics

Trends have the advantage of pushing brands to be as creative as possible while respecting a certain framework. Many good surprises can emerge from trends. For instance, the current trend toward a vintage look is still very much with us. And brands such as Tudor have become especially famous for excellent quality reinterpretations of historical models.

Two aspects are particularly relevant in ensuring the appeal of the watch will last: legibility and harmony. So, let's have a look.

Legibility

A watch is a small device, the goal of which is to provide the time quickly and easily. This means that you must be able to grasp the time in one glance. The key to achieving that is contrast. The hands, for instance, don't need to be fat or pink; they just need to be clearly localizable and recognizable. In fact, when you look at your watch, you quickly see the placement of two different-sized sticks and their positions in relation to each other. Of course, if you need to know the exact minute, it is necessary to have a closer look. But if not, a single glance should be enough. The same is true of digital displays.

To avoid reflection on the crystal, some brands use special antireflective coatings. So focus on that while you have the watch you desire on your wrist and ask yourself, "Can I read the time immediately?"

Harmony

"Harmony" is a magnificent word. It evokes peacefulness, balance, and a pleasing combination of those elements. Harmony works in the background to reach a superior state in which tensions don't exist anymore. Visually, harmony can be considered a combination of colors and forms that please the eye and calm the mind. On a dial, it can mean, for instance, that every element is clearly visible and at the same time fully participates in creating a pleasant general design. This is also true of three-dimensional parts such as the case that must have adequate proportions. Its lugs must make a fluid connection between the case and the strap. When you look for a watch, pay attention to subdials and chronograph counters, which shouldn't be too concentrated in the middle of the dial. Note also if some areas are unusually overcrowded with elements. Ideally, the information should be symmetrically located to create a general balance. And if you are not sure what harmony should look like in watchmaking, then take the time to really examine some classic models. You will see that their designs are, in the end, dramatically pure and clean.

Comfort

Sometimes there is a gap between something that "looks" good and something that actually "is" good. Watches are no exception. Certain models can capture attention but will be disappointing once they are on the wrist. If the model you like is uncomfortable, or if you can't close the strap without clipping your skin, or if the case is so big and heavy that you have marks on your arm, put it back down. And please don't fight the facts: they won't vanish. Quite the contrary! In the end they will divide you from your watch and you won't be happy to wear it. Comfort in a watch is as important as a good temperature for a bath or a wine that has no cork taste.

Two factors have a significant influence on comfort: size and ergonomics.

Size

The basic format of a watch comprises a round shape of a diameter between 39 and 43 millimeters for men and 36 and 41 millimeters for women. Of course, this is a baseline as no established rules exist concerning the form or size of a watch. What matters is that you feel at ease and comfortable with it. But it is good to take into account the ratio between your wrist and the size of your watch, especially with big and/or edgy models. When the watch is too large and sits too high, you may have some trouble pulling your sleeve over it. Tailors can help, of course, but is it necessary?

Ergonomics

An ergonomic watch will make you forget that you are wearing it. In other words, its shape makes it comfortable to put on the wrist and leave there for hours. We can define ergonomics by what it isn't. For instance, the back of the case or the lugs can sometimes be too sharp or edgy. Excess weight in an area of the case can create uncomfortable pressure. Pay attention to the strap or bracelet, too. Sometimes, for males especially, it is normal to lose some hair between the links of metal bracelets. Beyond this, a band and its buckle should fit your wrist perfectly and do absolutely no harm.

At the boutique, you may not have the time to fully test out the models so some problems may not be immediately perceptible. That doesn't matter. If you find a problem, the retailer should be able to fix it for you or propose an alternative.

Performance

A watch must be reliable and have good performance on several levels to be a true life companion. The most important factors are precision, water resistance, and dust and shock resistance.

Precision

What can we expect from a watch in terms of precision? How do we explain the paradox that a quartz watch's rate deviates infinitely less than its mechanical alter ego? With a few exceptions, the quest for superior precision is not a priority for most watch brands. Indeed, production processes today allow companies to manufacture watches, whose average level of precision can be considered acceptable. In fact, a watch's level of precision depends directly on the expectations put upon it. In other words, a deviation of one-thousandth of a second shouldn't prevent anybody from catching a train—or not.

In terms of precision, mechanical and quartz watches don't compete in the same category: thanks to their high-frequency regulators, quartz watches win the race by far. But mechanical watches honorably attempt the challenge; their performances in terms of precision are at least good enough to perfectly fulfill their roles. As far as rate deviation goes, quartz watches are less than a second off per day, while mechanical timepieces can be off by up to several seconds per day.

Historically, brands and watchmaking institutions began considering the precision expected from a wristwatch from about the second half of the twentieth century. At this time, electronic wristwatches were just being invented and their mechanical alter egos reigned supreme. It became necessary to guarantee minimal performance, which is why "norms" were prepared using tolerances defined by the watch-buying public itself. Based on the results, the COSC (official Swiss chronometer control) was created, which runs dedicated offices to test movements.

For mechanical movements, the protocol allows a timing difference ranging from 4 seconds slow to 6 seconds fast for calibers measuring more than two centimeters in diameter. For smaller movements, the range is -5 to +8 seconds. For electronic movements, the delta must be between -0.07 and +0.07 seconds per day at a temperature of 23°C, but extremely few quartz movements are actually tested.

Based on an international norm, COSC certification is still sought after today as it remains strong proof of efficiency and precision. Some brands certify almost all of their production, while others have decided to create their own protocols and test their products in-house.

"Chronograph" or "Chronometer"

There is some confusion about these two words, which deserves clarification. Many of your friends—and even the engineers who integrated the functions in your cell phone—use the term "chronometer" to describe a function that measures periods of time perhaps needed for a run or to boil a proper egg. Chronometer is a word that comes from ancient Greek and is formed by "chronos," meaning "time," and "metr(o)," meaning "measure." A chronometer, basically, describes any devices able to measure time. And, consequently, every watch is a chronometer.

But in watchmaking things are a little different: we use this word to describe watches with a superior level of precision certified by independent testing protocols. So only an especially accurate watch would be called a chronometer by these standards.

This mustn't be confused with "chronograph." In watchmaking, the chronograph is a function that measures short periods of time on demand. By default, a chronograph is a standard watch with this function added to it.

Water resistance

It was a long battle, but watches finally became water-resistant in the first half of the twentieth century. Today, all watches can actually be thought of as water-resistant except those with certain striking functions like minute repeaters as the vibrations of the chimes must sound through the case as strongly as possible.

ISO norm 22810 requires that a watch must be able to resist pressure of up to 2 bar to qualify as water-resistant. This is equivalent to a depth of 20 meters. However, most brands prefer a standard of 3 bar/30 meters.

Diver's watches obey the more imposing standards of ISO norm 6425. In this case, the depth must be at least 100 meters. In addition, the watch's most important criteria include being clearly legible and resistance to shocks, saltwater, and antimagnetic fields.

In any case, you still have to be careful with your watch around water. Even if the case has been thoroughly designed and built to prevent it, droplets of water and particles of dust occasionally find their way inside. This might happen if the crown has not been secured, for instance. But it can also happen if the watch's gaskets are too old or not put under enough pressure. It is very interesting that the deeper a watch descends in the water, the more all parts of the case will press against each other. Thus, as the watch goes deeper, it becomes more and more waterproof. In contrast, low external pressure can reduce the resistance of a watch, particularly when airborne in a plane as the air pressure is lower than on the ground. And, finally, it is during day-to-day life that a watch is most exposed to risks. Simply keeping your watch on your wrist while having a shower can be dangerous for it.

Consequently, it is a better habit to remove your watch when there are liquids around. And in conclusion, take note that the mixture of normal or low pressure with steam is the worst combination for your watch.

Shocks

Keep in mind that watches contain working parts that are practically invisible without a magnifying glass. This intrinsic fragility has occupied watchmakers and engineers for years, looking for ways to reinforce them and make the mechanisms stronger. Step by step, they have succeeded in inventing systems to protect watches from many of the usual shocks.

Today, watches can perfectly withstand ordinary shocks, but that doesn't mean they resist everything! Only certain models specially designed for the purpose can. And, strangely enough, the worst shocks are of the small, repetitive variety. Clapping your hands is especially famous as one of the

worst ways to treat a watch.
For the majority of timepieces, it remains necessary to pay attention not to smash them into things and not to wear them while playing intense sports like golf or tennis in which the arm is subjected to rapid accelerations.

ISO norm 1413 defines the shocks a certified watch must be able to handle. Testing for this norm includes two shocks, both of which reproduce an accidental fall from a height of one meter—the equivalent of the wrist of a standing person. The first shock occurs on the side of the watch, at 9 o'clock; the second is perpendicular to the crystal. In fact, the norm requires the use of a special tool that looks like a big hammer falling on the watch; it is impressive. The norm also stipulates that a watch must continue to work within a rate variation of a maximum of 2 seconds per day for mechanical watches and 60 seconds.

Antimagnetism

Strong magnetic fields can influence the accuracy of mechanical watches in particular. What happens is that magnetism disturbs the natural motion of the parts, in particular the balance wheel, the guardian of precision. Proximity to older electrical plants and devices emitting very strong magnetic fields (BlackBerries were famous for this) required an antimagnetic watch as a workaround. Antimagnetic watches have also been necessary instruments in certain industries or scientific fields. For instance, the first polar explorers needed this kind of watch for their missions since a magnetic network covers the globe with all lines converging at the north and south magnetic poles (which are not the same as the geographical poles). These two zones are the most magnetic places on earth. Polar researchers need special equipment to explore these areas.

Magnetic fields are still a concern today, especially because so many of our daily devices put out their own fields. You find them around speakers, cell phones, and computers. Watches are conceived to resist to them, but some can occasionally have a bad influence on a watch's rate.

Buying
Ten reasons
Choosing
Key elements
Aesthetics
Comfort
Performance
An investment?
The soul of the collector
Secondhand
Selling your watch
Knowledge

An investment?

Will a watch maintain its value over time? And asked in a more pragmatic way: can I expect some extra benefit or even profit if I sell it in the future? These questions have no easy answer. By default, and as with any consumer product, a watch loses part of its value the moment it leaves the store. Surprising? Let's remember that the same type of depreciation occurs with a new car. This is the general rule that concerns almost all watches currently in circulation. It is perhaps wise to learn whether one can financially speculate with watches as an investment model.

Nevertheless, more and more watches are going for millions at auction today. Given that, you may ask, can my watch be considered an investment item? The answer to this question is subject to a lot of variables, but we can focus on a few elements. In any case, this is a discipline that only a small group of professionals really master. Similar to wines or cars, certain brands exhibit superior strength in the secondhand market—these are Patek Philippe and Rolex in the watch world, which owe this to their rich history, very well-executed models, strong reputations, and the fact that they are financially independent companies. Furthermore, they became strong because of their success among collectors—some of whom are extremely rich and able to bid fortunes to buy a particular model.

But not even every model from these brands can become "unicorns." Those that do must have something really special, be exceedingly rare, and in the best condition (preferably unworn). They are also often unique pieces, a very important point. For instance, a prototype, a dial bearing a mistake, a watch owned by a celebrity, or an example from a very restricted series can cause the selling price to explode.

The soul of the collector

Sometimes a simple yen for watches can take on huge proportions. It can become so absorbing for the passionate that some of them mutate into hunters tirelessly chasing their prey all over the world. They spend all their time on sales platforms and carefully watching auctions. Competition heats up among them as they all run for the same models. But this intensity only concerns a minority, and most collectors live their passion quietly.

Building a collection is like creating a story of which you are the author. There are many ways of framing a collection, but the majority of collectors focus on one brand. Some choose to buy what they like. Others focus on certain complications. Others pick models produced during a certain period of time. Some look only for customized versions. There are no rules outside of the fact that you must like it. A collection is really a personal expression. And it also naturally evolves throughout time. Many collectors decide to sell their watches to buy the next ones. If you are friendly with this kind of collector, it can be a great opportunity to get a special deal.

Secondhand

The secondhand market grows every day and offers really great opportunities. In addition to traditional sellers and auction houses, dedicated websites have popped up and drive a lot of business in secure conditions. It can be like a jungle out there, though, and the key factor is knowing your retailers enough to trust them entirely. The longevity of the company, its size, the quality of the service, and the functionalities of the web platform are some signs that indicate professionalism in the way they do business.

In any case, never dither with the provenance of the watch: it must be crystal clear. Serious sellers provide an official and detailed invoice with all the papers concerning a given watch. Every original document is valuable, as well as every item linked to the watch—box and straps, for instance. These can also increase the value of the watch if you want to sell it later.

Be aware that watches may have been modified throughout time. Parts may have had to be changed while others may have been customized, even within the movement. Naturally, it is very difficult for beginners to establish this, and even experts sometimes meet with great difficulties in identifying them. The best thing to remember is that the more original and in good condition the watch is, the higher the price will be.

The brands themselves may be your best allies in confirming authenticity. So try to curb your enthusiasm and go for deeper analysis if possible.

Individuals

Many secondhand watches are exchanged directly among friends or aficionados. Try to find a group in your area and get in touch with its members. With such groups it is first a question of passion, so they should welcome you easily and help you with advice and contacts.

Secondhand shops

Secondhand shops are radically transforming from the messy, dusty old stores they could have used to be into modern and superefficient boutiques. The level of expertise in secondhand watches has increased significantly, and connoisseur-entrepreneurs have started making money out of it. Secondhand shops are particularly attractive because most of them offer watches at a multitude of budget levels. In addition, their watches will perhaps have been serviced and repaired and sold with a guarantee. Professional secondhand shops are great places for watch lovers who want to get into collecting.

Auction houses

In the main auction houses, most of the watch departments have been expanding for about thirty years. Their business model is to sell watches that have been entrusted to them to obtain the best price during auctions with buyers present in the room from all over the world or bidding by phone or Internet. And as some collectors really spend millions for a single wristwatch, such events are always widely covered in the press.

More affordable models are included in these sales as well. Before the sale, the auction houses publish a catalogue with all the models included and their price estimations. Try to get one if you are motivated. And if you don't want to spend money on the catalogues, you can see them online and also visit the exhibitions where all the participating models are presented.

Online

Watch retailers and brands vacillated for years about selling watches online. Those days are over, and solid platforms have emerged. They provide the necessary guarantees—including complete profiles of sellers, all the specifications of the watch, and of course secured payment. Retailers and secondhand shops also show watches in stock online as a matter of course these days. Thus you can find great deals on the Internet. But not every watch for sale is visible online. So continue to visit brick-and-mortar boutiques regularly, explore new ones, and talk, talk, talk with other collectors.

Authentication

It boils down to this: only a brand has the legitimacy and authority to authenticate a wristwatch. But this hinges on the brand's desire to do so, of course. A retailer can give an opinion, but without papers or in case of doubt, only the brand can preside. This process can take time and costs some money.

Warning: no stolen or fake watches!

You can be in serious trouble if you buy a watch that has been stolen or is a fake. Also, a watch that has been lost and not given back to the owner or to the police is considered stolen even if you have legitimately purchased it. So be on your guard and always ask for proof regarding the origins and authenticity of a watch. If you find yourself seriously attracted to a fake watch, hold on and think a little further. By buying a fake watch you are betraying yourself and the people in the industry who make their livings on original work. In addition, even if they are really well done, fake watches don't have the same guarantee of quality that real ones do. Also forget about any official help or after-sales service or thinking that you are making a good deal as you can still spend several hundred dollars for a single one. Once the watch dies, it's over. And they sometimes die very fast.

Fake watches are also illegal. So if you get caught crossing a border, you lose the watch and gain a fine—one that you generally will remember.

This is also not only about ethical, quality, and financial concerns. The fake watch trade is controlled by organized crime, which also deals in drugs, prostitution, guns, organ trafficking, etc. These groups exploit adults and children under ignoble conditions. Fake watches belong to a poisonous economy. By buying a fake watch, you are contributing to these businesses' health and longevity.

Selling your watch

You have experienced good moments with your watch, but you have decided it is now time to let it go. How should you go about this, and how much can you ask for it? Only experts can give you an adequate estimation for your watch.

The most important criteria are the general state of the watch and if it is working. So you're better off having it serviced first. Official papers are important, too, as are the original invoice and guarantee. They are proof of the origin of the product. In addition, aficionados are always interested in archives, packages, replacement straps, and even gifts that came with the watch. They amplify its historical value. If you don't have any proof of origin, only the brand can certify it. Depending on the model, it may be worth making the effort—in terms of time and finances—because certification can really boost the price.

Once the watch is in the state you want for selling it, you have to fix a price. If you contacted an expert, he or she will already have given you numbers. If you didn't, then you should find one who will provide this advice. Usually, a good way to go about this is to ask him or her to sell it for you in exchange for a commission. You may not have the network to generate enough interest and so selling it yourself could take a long time.

Knowledge

Do you appreciate watches, but don't have a clue about them? To continue your exploration with pleasure, you have to start building yourself a small watch culture. Media gather a lot of information and are easily accessible. Those dedicated exclusively to watch-making, whether digital or not, offer a range from the worst to the very best quality in content, like blogs. You have to persevere to find the good ones: they exist! Video content about watchmaking is still in its baby shoes, but it is a very promising sort of media as animated images are very expressive. Of course, there are also books. Unfortunately, most of them are too specialized and dedicated to offering professional information, or they were made at the request of a brand and their pertinence must be evaluated with a certain critical approach. You get the point: building a watch culture is a pathway without a preestablished plan.

Internet

It is so easy to create a website that plenty of them have popped up. Among them you will have good ones and useless ones. To identify the real players, you should consider if they frequently publish original and varied stories. Try also to evaluate if they are well connected to the watch community by liking their fan pages on Facebook, for instance. Brands regularly send press releases to the media, and some websites just publish those—which is not so interesting as they don't add any personal input. Key players are the ones showing a real understanding of the products and the brands, making a sincere effort to share their interest as frequently as possible.

Media

Traditional media usually employ their own dedicated journalists. Their opinions are valuable; indeed, they are more independent in their way of commenting on what they see. They are often opinion leaders, too. Some of these media outlets also work with external independent journalists. If you liked an article by one of them, remember the name and do an online search to access more stories by those individuals.

Blogs

Dedicated blogs are very active in watchmaking, too. Many have flourished thanks to young enthusiasts who wanted to share their thoughts about what they love in a totally free way. Naturally, the early days could be chaotic for some of them, and the quality and accuracy of the content is not always relevant or right. But some of the ones who have stuck with it have become serious players with strong credibility. The major difference between blogs and traditional media is that blogs have remained consistent in their core business: sharing personal views above all. As for journalists, the best idea is to find the ones you like reading and follow them.

Institutions

Museums, foundations, and associations also provide interesting information. As they are generally entirely dedicated to their fields, the quality is good. This positive aspect can also be a limit as you are entering a world of "geeks." Topics are very precise and investigations very deep. It is not about trends or the reputations of brands here.

Watch lovers

Involved watch lovers are likely to meet online. Some are active on forums that are actually communities set up to discuss new watches, technology, history, and the best deals all day long. Often a group of people post comments or start discussions. The quality and level of interest can vary a lot, but there will always be a bunch of big fish who really know a lot. Try to identify them and focus on their interventions. You will learn fast and well. Of course, you can ask questions, and you should get answers. If not, someone will indicate to you for sure where you can find them.

How to find an expert

Watches have been sold worldwide since at least the sixteenth century. This means that you can find expertise in any region of the globe. The first door to knock on should be one of a well-established retailer selling various brands. Such vendors combine experience and the ability to deal with different kinds of products every day. If you are not satisfied with the answers or service you get, do your homework and identify auction houses, commercial agents, or influencers and get in touch with them.

Care

Congratulations, you just purchased your watch.

Now take your time and enjoy the delicious pleasure of wearing it as often as you want. Today any “serious” watch is a solid and reliable instrument. But this doesn’t mean it is indestructible. In spite of appearances, keep in mind that a watch remains a delicate and refined object.

Even setting the indications on your watch is not a harmless act and needs some practice. Servicing it is also important to ensuring good operation.

Below you'll find some key points that you need to know.

Setting

Basic functions are set using the crown, which works as a selector that provides access to the hours and minutes, calendar indications, and even the moon phase indication. It also allows the wearer to wind the movement, even on automatic models. On complicated watches, not every function can be set using the crown; correctors or pushers—or both—might be necessary. If your watch in particular has correctors, please use the appropriate stylus tool to activate them. Usually, this is a strong nonmetallic pin that is delivered with the watch. The danger here is in scratching the case if you don't use the proper tool.

Setting can be delicate even on simple watches because it momentarily disturbs the usual running of the movement and requires parts to work differently than they normally do. Depending on the function, you may be directly accessing an internal system in action, like the one that changes the day, for instance. On most watches it is advisable not to adjust the date when the hands are between 10 and 2 o'clock—when the date mechanism is in the middle of changing—to avoid damage.
In addition to that, some functions can only be set by turning the crown in one direction. Turning it in the opposite direction might break it.

For a few years now, brands have worked hard on securing movements in order to prevent damage while they are being set. But such safeguards have not become general at all.
That is the reason why each watch must be always carefully handled—especially old ones.

You should be able to set your watch yourself. Your user guide normally explicitly explains every step. But if something is not clear enough, your retailer will know how to proceed.
A friend with experience can also be helpful. Some complicated watches, such as perpetual calendars, for instance, require some experience to set because of the intrinsic mechanism. If you have any doubt, hold on and find a real expert. Some videos available on the web can help. Before following any advice you find online, check the validity of the authors. And never forget: if something happens, you and only you bear responsibility for the problem.

Maintenance and care

You can clean your watch yourself if you follow some rules and apply some care. In any case, a soft piece of fabric makes for a good ally for a quick cleaning. A soft toothbrush and normal soap are usually enough to remove dust and impurities on waterproof models with metal bracelets. And for heaven's sake, don't clean vintage watches yourself or try to use soap and water on leather straps! If in doubt, ask a specialist! Leather doesn't stand humidity at all. Ignoring this warning reduces its life's expectancy.

Dry the watch afterward with a soft cloth. Of course, adapt to the delicacy of the task at hand: don't rub a watch hard as if it were a pan in which a steak has burned. Exhibit a lot of love and sensitivity. And above all, don't be too harsh with the moving parts of the case such as the crown and the pushers. If you don't feel confident with this operation, ask a professional. Retailers are equipped with special equipment and ultrasound systems especially conceived for this task.

Of course, basic cleaning won't remove scratches and other marks. The watch must be polished by a professional to remove those. This is an operation that should be undertaken exclusively by professionals as you could seriously damage your watch if you are not experienced. Consider also that in addition to metal watches, plastic ones can also shine again. A good polishing is almost free for cheap plastic watches but can cost hundreds of dollars for a prestigious metal timepiece.

Watchmakers are of the opinion that a new watch can be worn between three and five years before you need to have it serviced. Others say that you don't have to touch it before it stops working correctly. Actually, both options can be correct. Older watches need special care, and it is better to anticipate them breaking by regularly having them serviced. In such cases, parts can't usually be replaced or repaired; they must be totally reproduced, which is a very important, costly, and time-consuming job.

Service and after-sales

As with a car, a watch must be serviced from time to time. The brand or the retailer will tell you when it is best to come back. Be aware that brands run wide after-sale services networks with specially trained specialists. It is better to take advantage of this service more often than not.

Repairs

If your watch has any trouble, don't wait to contact customer service or your retailer. You can do serious damage to your watch if, for instance, it is no longer water-resistant, which might be the case following a shock, or if the oils that help lubricate your caliber dry up and become sticky because of dust and impurities. Oils can also evaporate. If that happens, there is much more work to do, and the costs will increase proportionately.

You might be surprised at how high the costs for repairs can be. In addition, you may have to wait several weeks (or months) before you get your watch back. This is how watchmaking brands mainly work today. To make things go easier and faster, you can try to find an independent watchmaker to fix your watch. But this is at your own risk. First, watchmakers able to do things properly are rare. And they are often quite busy, which means that they will need time to work on your piece. Second, spare parts are becoming more and more difficult to find, and their prices match their rarity. Third, any official guarantee will be automatically canceled the moment the watch is touched by a third party. And, finally, you have a guarantee that the job has been well done only if the brand, a certified retailer, or an authorized after-sales service handles it.

Travel

This is not rocket science: you should be aware, really aware, when you are walking across a train station or going to your gate at the airport that pickpockets like to take advantage of travelers' inattention or fatigue.

Less dangerous but still highly annoying is when you forget your watch on the nightstand table at the hotel, in the washroom next to the soap, or at the airport's security check. These are true stories.

At security control, you might have to actually remove your watch to pass through the electronic sensor. Don't be afraid of letting it go through the hand luggage scanner: X-rays have no impact on it.

In a plane, be particularly careful if any liquid touches your watch. Because of the lower air pressure, the case will be slightly less water-resistant.

It is helpful to bring a dedicated watch case with you when you travel. This way, you have a protective case that can be practical for many other occasions, too. If you travel with several watches, you can certainly find larger ones.

Insurance

Basic insurance may already provide some protection for your budding watch collection, but before signing up, check it out carefully. Good coverage can very be helpful in case of trouble—in particular if your watch gets stolen or if you lose it. As prices and conditions vary a lot depending on the watch itself and the region where you live, it is best to ask your retailer and your insurance provider.

Icons of the twentieth century

Certain watch models have become so successful over time that they are now considered as references in modern watchmaking. Others combine many major characteristics to distinguish themselves, even if they have not entered into the legend category with the same aura as their peers.

The following selection brings together eleven models that can be considered among the most interesting today. For the majority, fortune came with time. No one can predict if a model will be authoritative the day it is launched. And professionals say that it takes at least ten to fifteen years before we can gauge the kind of success that makes a legend. Finally, even if many of these models cost a certain amount of money, others remain easily affordable.

Audemars Piguet
"Royal Oak"

In addition to a revolutionary design for that time, the *Royal Oak* broke a taboo at its launch in 1972: Even though it was made out of steel, it would be sold for the price of a luxury watch.

AP
AUDEMARS PIGUET
8

CHRONOMETRE
12

Breitling
"Navitimer"

Available since 1954, the Breitling *Navitimer* is a chronograph with a rotating bezel. It works as a slide rule and was developed especially for aeronautical navigation.

Cartier
"Tank"

Launched in 1919 and designed by Louis Cartier himself, the *Tank* was inspired by the shape of those armored vehicles. It is one of the oldest wristwatch models on the market.

CARTIER

REVERSO

Jaeger-LeCoultre
"Reverso"

In 1931, Jaeger-LeCoultre introduced the *Reverso* with an articulated case that allows it to flip along its vertical axis in order to protect the dial, for instance while playing sports.

Longines
"The Longines Master Collection"
(ref. L2.673.4.78.3)

The long and rich history of Longines allows the brand to offer high-quality and well-executed models for affordable prices.

MON
JAN
LONGINES
AUTOMATIC

PATEK PHILIPPE
GENEVE
18
SWISS
MADE

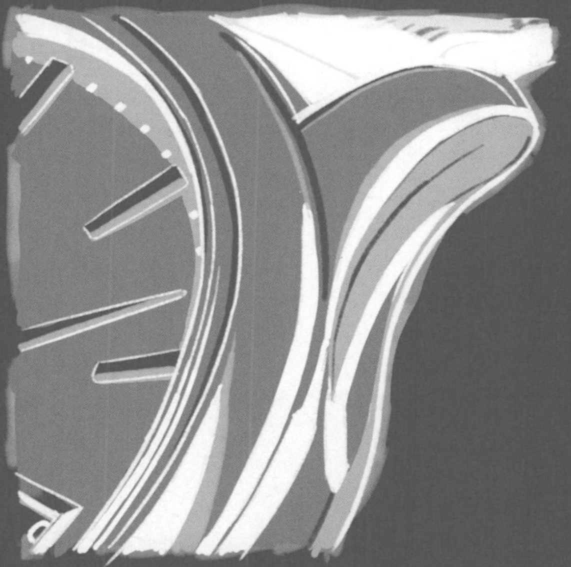

Patek Philippe
"Calatrava"

Introduced in 1932, the *Calatrava* from Patek Philippe has become the emblematic model of this Geneva-based manufacture. Its design has subtly evolved throughout its history.

Omega
"Speedmaster"

Nicknamed the *Moonwatch*, the Omega *Speedmaster* was the first wristwatch ever worn on the surface of the moon. It happened to be on the wrists of Neil Armstrong and Buzz Aldrin on July 21, 1969.

OMEGA
Speedmaster
60
40
20
30
20
10
12
9
3
6

50
10
40
20
30
ROLEX
SUBMARINER
SWISS MADE

Rolex
"Submariner"

Conceived for underwater activities, the *Submariner* model by Rolex was launched in 1953. Since then, it has become a watchmaking must-have because of its reliability, performance, and sporty-chic look.

Swatch
"Bellamy"

The launch in 1983 of the brand Swatch marked a significant moment for the Swiss brands. Thanks to its entry-level plastic models, it played a decisive role in the rebirth of the Swiss watch industry.

swatch
SWISS

MONACO
HEUER
8

TAG Heuer
“Monaco”

With its square case, the *Monaco* shows off a rare form that succeeded in becoming a reference among wristwatches.

Zenith

"El Primero"

In 1969, Zenith presents the *El Primero* chronograph, meaning *The first* in Esperanto. It became famous because of its performances at the high frequency of 36,000 beats per hour with automatic winding, a revolution at this time. It disappeared from the market shortly after being launched for industrial and economic reasons. But it reappeared almost miraculously in 1984 to become, since then, one of the references among mechanical chronographs.

ZENITH
36 000 VpH
TACHYMETER

Anatomy

A watch is composed of dozens of components. We will look at the principal parts within the technical organs they constitute in the following pages.

Watchmakers make a distinction between the mechanism itself and all other parts, in particular the exterior, visible ones. The former deal with time measurement and functions, while the latter relate to legibility, aesthetics, and comfort. In short, these second ones “dress” the watch.

Display

The display of the watch works like a monitor. It shares the information produced by the movement or the chip. All major indications are related to timekeeping, but super-rare functions, on mechanical watches at least, like a thermometer or pedometer (the measure of distance in steps) can also be included but are considered secondary. From the traditional counter to the digital screen, there are a lot of options. Take your time to find out which one you prefer because there are huge differences.

Traditional dials are made of solid materials and can be luxuriously decorated. On the other hand, digital displays work like screens, are made exclusively by machine, and need electricity to function.

Fig. 01: *Dial*

Fig. 02: *Screen*

Dial

The dial is the “face” of the watch. So it’s worth paying close attention to how it is designed and constructed. Fine mechanical timepieces and electronic watches have almost nothing in common; they encompass two different worlds entirely. But the final objective remains the same: you must like it and be able to read it at first glance. So what is a good dial, and how can you recognize one?

Screens

The traditional dial transforms into a screen on electronic watches. So after confirming it works properly, the only other verification is to be sure of its quality. Generally, firms that produce these watches also learned how to produce their own.

In 1970, the first electronic watches used LEDs (light-emitting diodes) to display the time. Later came the LCDs (liquid crystal displays), which now constitute the standard. This system uses liquid crystal, which can stop light or let it through thanks to electrodes. The dial, run by a microprocessor, can display all forms of numbers and indications. Today, some LCDs are backlit with LEDs. Recently, OLED technology has popped up. OLEDs—organic light-emitting diodes—are particularly interesting because of their low energy consumption and can be fixed on flexible surfaces.

Sometimes, a watch screen may look like a Christmas tree at night.

Anatomy
Display
Dial
Screens
Traditional dials
No dial
Indications
Functions and complications
Case
Straps and bracelets
Materials
Movement

Engineers and designers can indeed add so many little windows that the ensemble looks like the control board of a Star Destroyer. If you like this kind of display, just go for it! But remember that the more confusion you find on this fundamental element, the less you'll appreciate using your new toy.

Traditional dials

Quartz and mechanical watches have solid dials manufactured through various operations. Almost everything is possible in terms of creativity, materials, finishing, and colors. Production processes can be specific, complex, long, and sometimes they are only possible with handwork. Also, they can be considered testimonials for what lies behind them: the quality of the movement and its features.

The common dial is flat. It is the most basic and frequently used, but it is not the only option. Some dials also have holes, called apertures, to let the owner see one or more so-called disks situated underneath. Disks show indications such as the date or moon phase.

Domed or curved

Dials can be domed or curved. These designs create visual effects or play with proportions and volumes. Domed dials are usually shaped this way to better fit the movement and make the watch thinner. Pay attention to one detail in particular with domed dials: the shape of the hands should follow the curvature of the dial. Very often these evoke vintage looks from the 1940s and '50s because domed dials constituted a trend at that time.

Multilevel
Dials can have several levels. This means that the surface is no longer flat like a coin but that some parts are lower or higher than others. These dials are more complex to design and build, but there is a purpose: to strengthen the personality of the watch and also aid readability. This concept is often used to good advantage to distinguish different kinds of indications. For instance, the counters of a chronograph might be positioned slightly below the main dial to signify that they have a different mission. In addition, because the dial is composed of various parts, designers use decoration or finishing that would have been harder to create in a one-level dial.

Disk
A disk is a mobile dial of reduced size revolving underneath the main dial. Usually employed to display the date, the day, or the phases of the moon, disks become visible thanks to an aperture cut into the main dial.

Openworked
Dials can be sometimes described as "openworked." This means that windows have been cut that allow a view all the way through to the movement. This constitutes one particular way of decorating the dial of a watch.

Fig. 03: *Domed or curved dial*

Fig. 04: *Dial with multiple levels*

It can also expose a special technical aspect of the movement. For instance, Frédérique Constant has a "Heart Beat" collection whose models feature an aperture above the escapement, a crucial mechanism for every mechanical movement.

No dial

What about removing the dial altogether? To offer clear insight on the workings of the movement, some watches have no dial at all. The caliber itself becomes the point of interest: its shape, design, and the color of the various components catch the eye. The choice to have no dial mainly strengthens the technical character of a watch. On high-end mechanical watches, the aim is to show the finishing and fine decorations on each part and to reveal the architecture of the movement.

Indications

Watchmakers have developed various techniques to display all available indications. It is a huge challenge to succeed in making them readable, balanced, and beautiful all at the same time. Thus, a dial can contain a multitude of scales, dots, numerals, markers, and graphic "tricks" to allow a maximum of the options at their disposal. Indications are usually printed. But they can also be three-dimensional. Mostly used with luxury models, these are "applied" (they are sometimes called "appliques") to the dial on little pins that are riveted through micro holes in the dial.

Reading indications

Some dials display multiple indications that sometimes look messy and confusing at first glance. Take the chronograph, for instance. If you are not used to reading one, it might be hard to immediately understand the role of each counter and scale.

Watchmakers often use colors, numbers, and graphic codes to create distinctions in a rigorous and coherent way. Consider colors, for example: all the indicators (hands, dials, subdials, and scales) belonging to one function should share the same color. So when you look at a dial, pay attention to visual repetition, the scale of every counter, the shape of the hands, the fonts, etc.

Fig. 05: *Indexes*

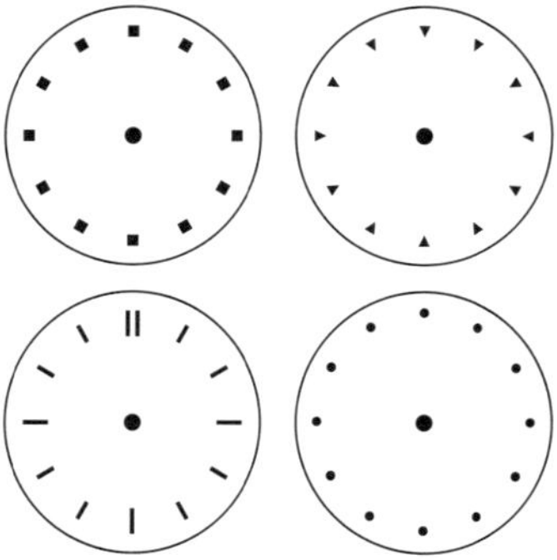

Indexes

Indexes are tiny parts on the dial usually made out of metal and sometimes covered with luminous material, which means they shine in the dark. Precious stones like diamonds are frequently used for markers on luxurious watches (mainly women's models). The main goal of the indexes and markers is to indicate clearly a 12-hour scale. Indexes can have plenty of forms: rectangles, triangles, or dots, for instance. If they are numbers, then an Arabic or Roman font is used. But in reality, there is no strict rule here. Indexes can also be completely removed from the dial. In 1947, American industrial designer Nathan George Horwitt conceived the so-called "Movado Museum Watch," a timepiece with no indexes at all, just a large dot at 12 o'clock to represent the sun. Horwitt used the fact that human beings had become used to reading the time by the positioning of the hands and the space that separates them to his advantage. It became an icon of horological design and is still in production.

Symbols (logos)

The logo of a brand can be considered its visual signature. Usually positioned at 12 o'clock, it can be very simple and stand alone or it can be positioned along with the name of the brand. In very rare cases, neither logo nor name are visible on the dial; brands that go this route count on reputation and recognizable design codes to distinguish themselves

Fig. 06: *Minute scale*

Fig. 07: *Sector scale*

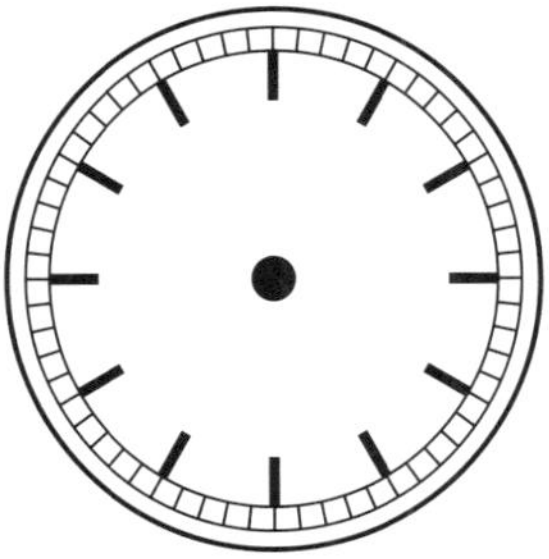

Fig. 08: *Tachymeter*

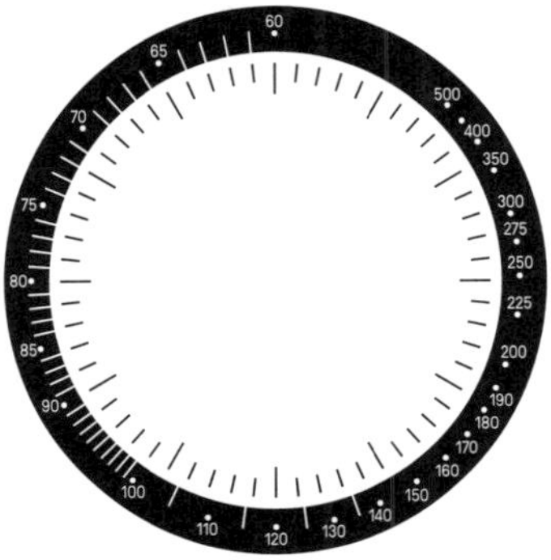

from others. In 2014, De Bethune, for instance, launched its "DB29 Maxichrono Tourbillon" with neither name nor logo on the dial. Another option is exercised by the familiar Bulgari brand, which is engraved twice on the bezel of the "Bvlgari Bvlgari" model. Gucci makes some of its watch cases in the shape of a G.

Scales

Scales are necessary to precisely display certain information like the legibility of minutes and even seconds, which is important on certain models—chronographs in particular. Some scales use existing indications to offer additional information. Such is the case for the tachymeter scale that measures the speed of an object using the chronograph's second hand with a dedicated scale generally visible on the bezel or around the flange.

Minute scale

The minute scale is the most common scale visible around the circumference of the dial. This is the obvious design choice when it is necessary to read the exact minute or second.

Sector scale / "Railroad track"

When the minute scale is made out of two concentric circles connected by 60 cross lines, it is called a railroad track or sector scale.

Tachymeter

A tachymeter is employed to determine the speed of an object like a vehicle or a race participant.

Anatomy

Its scale is generally located on the bezel, like on the legendary Omega "Speedmaster," where it has become one of the watch's distinctive features. A tachymeter requires a chronograph function because the user needs to be able to start and stop the second hand on demand. The measurement stops once the object has gone 1,000 meters or feet, depending on the choice of the measuring unit employed. The second hand automatically displays the average speed on the tachymeter scale.

Pulsometer

A pulsometer scale allows the watch's wearer to calculate heart rate directly from the displays on the dial thanks to a chronograph version. Pulsometer dials are usually calibrated to 15 or 30 pulsations. This means that the nurse or the doctor stops the measurement after 15 or 30 beats to read the quantity of beats per minute directly on the scale.

Telemeter

Thanks to the chronograph function again, a telemeter scale allows its user to determine the distance between the place where the watch is and a visible event. One clear example is to determine the distance of lightning during a storm. To do this, start the chronograph when you see the event occur (lightning flash), and stop it when you hear the sound (thunder). Then you can read the distance on the dial's scale.

Fig. 09: *Pulsometer*

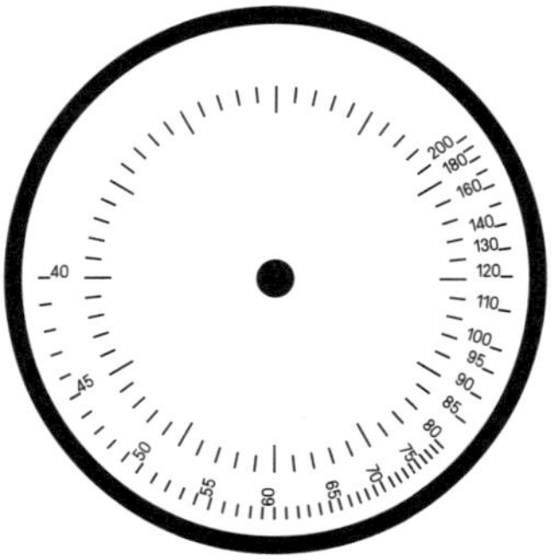

Fig. 10: *Telemeter*

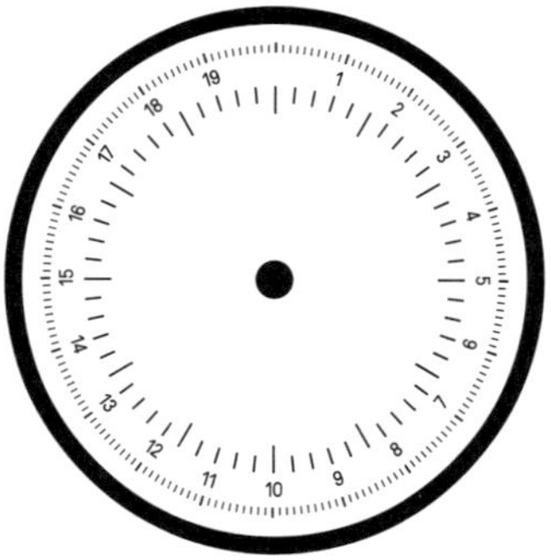

Dial decorations

A dial's decoration is of utmost importance as the dial is the "face" of a watch. With options ranging from a simple color to a complexly engraved pattern and galvanic coatings—and even more refined decoration techniques—dials can be considered as if they were little pictures. The work that goes into them is even more impressive when you understand that they must be perfectly reproducible in a serial manner.

Lacquer and coatings

Different materials and techniques exist to color a dial and protect it from fading. The most common of these is lacquer, which allows all kinds of variations. If the structure of the dial underneath must remain discernible, coatings are used instead. Transparent or colored, these don't hide the surface of the dial.

Structured

Dials sometimes have patterns impressed into them, which are either stamped or produced using an engraving technique called guilloché. This means that the metal has been sculpted using a variety of techniques to create forms and visual effects. These designs are produced either by machines or hand-driven tools. Dials decorated by hand are particularly valuable, and some patterns act as distinctive visual signatures for watches. The casual sports model by Audemars Piguet known as the "Royal Oak" would be one example. Indeed, its "tapisserie" (tapestry) dial pattern constitutes one of the major signifiers of the collection's identity (see p. 82).

Stamped decorations

Like a coin, a dial can be stamped to create a pattern on its surface. A raw metal blank is fixed on a support. Then a metal die presses into metal, deforming it and impressing the design onto it. This technique is most commonly used for large series and gives very good results. You need to be a real expert to notice its use in place of handmade decoration.

Guillochage

Guilloché is a type of artisanal delicacy comprising a repetitive engraved pattern executed with a hand-driven machining tool. Also employed in jewelry, it makes symmetrical and geometrical patterns in the dial. The name comes from the French "guilloché," which refers to the technique of "guillochage." The artisan using this hand-driven machine creates and reproduces patterns. The technique also allows several kinds of patterns to be engraved on the same dial, a feat that requires rare know-how as the craftsman has to "program" the machine by changing out and even making the cams needed. An alternate way of creating guilloché on a dial is by using a pantograph. This mechanism actually "reads" the pattern on a master as it automatically reproduces it on the dial.

Both of these techniques have been used for centuries but almost

Anatomy

disappeared in the last thirty years. Thanks to the revival of high-end watchmaking, they have been saved. Much effort has been put into creating new patterns and extending the possibilities of guilloché.

Decoration patterns

Straight lines, diamond shapes, and intersecting square patterns are the most common dial decorations used in guilloché. Note that the diamond shape has its own name: "barleycorn." A square pattern reproducing a small pyramid is called "clous de Paris." Shown here are a few examples of patterns.

Fig. 11: *Clous de Paris*

Fig. 12: *"Tapisserie"*

Fig. 13: *Sunburst*

Fig. 14: *Barleycorn*

Fig. 15: *Concentric*

Fig. 16: *Lines*

Crafted dials

Some dials have another mission beyond that of a refined space for indications. This little surface may indeed become a piece of art. Techniques like engraving, enameling, micro painting, micro sculpting, and gem setting can enrich the dial's surface. The techniques involved are essentially traditional and must be almost exclusively executed by hand. By definition, these dials are unique pieces, costing many hours of work, with the price of the watch in line with the effort.

Different levels of decoration exist for artistic and crafted dials. These can range from very pure and uniform to highly varied. Nevertheless, even the simplest already boasts a higher cost due to the rare technique and the specialized know-how involved.

— Engraving

Engraving is another artistic technique. Various types exist that can be used individually, combined with others, or adapted following a theme in order to decorate components with motifs or patterns. Engraving can also underscore certain details and allow for personalization of the watch.

— Gem setting

Dials are sometimes partially or completely set with precious stones. When the dial is entirely covered with gemstones, it is described as “fully paved” (from the French “pavé”). The arrangement of stones naturally follows a pattern to create visual effects. Different kinds of settings exist that depend on the shape of the stones as well as the type of effect that is desired.

— Miniature painting

Miniature painting is the art of executing minuscule images on a dial. Historically, craftspeople used to reproduce classical paintings using a very delicate enamel technique. Today, enameling is still popular for its intensity of color and longevity, but most "miniature" dials are painted using synthetic paints.

Dial materials

Dials are usually made out of brass. Precious metals such as gold are used for very exclusive timepieces. Carbon fiber, ceramic, and silicon are employed as technical alternatives. Watchmakers like using sapphire crystal to obtain transparent dials.

Refined dials

Generally, precious materials are used to create artistic dials because of their interesting technical properties. Gold, for instance, is by far the best metal for engraving and gem setting. In other cases, the prestige and coherence of the whole product will determine the choice of material.

Precious metals

Any kind of gold, platinum, palladium, and even silver can be used for dials. Beyond their rarity, each brings a specific color and shine to the dial, offering great results for certain types of finishing such as "grainage."

Precious and rare stones

Precious and semiprecious stones naturally find homes in luxury watches. Here you can find dial plates made of opal, turquoise, lapis lazuli, tiger's eye, natural crystal, jade, malachite, and different types of ruby. A dial usually measures less than one millimeter in height!

Sometimes extraterrestrial stones are used—meteorite being the most common. Their origins in outer space make them absolutely unique, and some have particularly interesting structures. Such dials can be challenging for watchmakers, though. Despite the best machining and polishing techniques, they may still be too thick and slightly irregular. One of the most interesting skills experienced watchmakers have is to be able to adapt the inside of a watch to receive this kind of dial.

Mother-of-pearl

Mother-of-pearl comes from oyster shells usually employed for the production of pearls. Disks are cut out of the shell before leveling. Natural mother-of-pearl can come in shades of white, blue, yellow, brown, and black, but this material can be artificially colored as well. In any case, mother-of-pearl dials are pleasing because of their brilliance and the way the shimmering material evokes clouds, the surface of water, or fields of snow.

Enamel

Enamel consists of crushed glass colored with metallic oxides. Each mixture is diluted in a little bit of distilled water and applied to a dial with a very tiny brush to create an image, a pattern, or to fill prepared cavities. The artisan uses only one color at a time, which is fixed by firing in a kiln before the next one is applied. Firing transforms the powder back into glass: the dial is put into a special oven at a temperature of more than 800°C. No one can predict if bubbles or cracks will appear at each firing, and such imperfections result in the dial being discarded as damaged. Just imagine: Some executions require several weeks of work. If a flaw shows up close to the end, the loss is huge, and the artisan must start all over again. This is why the best enamel specialists are rare, very gifted, and have terrific patience.

The most common technique used for "haute horlogerie" watches in Switzerland is called "grand feu technique de Genève" as it involves a firing process. If a pattern is involved that is created by engraving the metal dial first to form cavities filled with enamel, it is called champlevé. The better known cloisonné process sees sections created by welding an ultrathin strip of metal, usually gold, onto the flat dial before the enameller fills them one by one with enamel. Paillonné enamel includes tiny motifs stamped in precious metal—gold most often—added into layers of enamel to create a decorative pattern.

Finally, enamel can also be used like paint to reproduce images; this is called the miniature enamel technique. More demanding by far and requiring real artistic skills, oil is used here instead of water to fix each color. The profoundness, the thinness, the intensity, and the longevity of such realizations make this technique exceptional and without equivalent.

Display modes

On watches, hands do the heavy lifting to display the time. But for electronic timepieces, a screen reproduces artificial hands or the time is given digitally. A very few intrepid watchmakers have also chosen to use moving ribbons or rolls. In this supreme category, even liquids and magnetic systems have been employed for the most disruptive of luxury watches. Some models also combine different kinds of displays in order to improve the legibility of the watch. Whatever you choose, be sure to select the most comfortable and easy solution to look at.

Fig. 17: *Analog*

Fig. 18: *Digital*

Analog
When hands are used to indicate the time, this is an analog display. The hands don't show the time themselves, they point at a number or a zone on the dial. This is the less intuitive method: remember when you were young and had to train yourself to read a clock? Still, the analog display remains the more common type.

Hands are delicate parts that can be given different shapes and colors to convey a specific look.

Digital
A digital watch displays the time using numbers. The digits are there in front of you for reading, no interpretation required. The digital mode of display is the most common among electronic watches. Some mechanical watches have also been conceived to indicate time this way. Paradoxically, such movements are among the most difficult to create because of the forces needed to start and stop the disks upon which the numbers are printed. At the extreme end of this idea, the de Grisogono "Meccanico DG" mechanically controls seven segments needed to form digital numerals.

Indications
From an immaculate surface to a composition of lines, numbers, and words, dials can go from pristine to balancing a variety of indications. In the most common configuration, twelve numerals are placed around

the circumference starting at the zenith of the dial. Somewhere in the middle or around this circle the name of the brand and potentially the country of production appear. On older timepieces, the number of jewels used as bearings was also noted as a distinctive feature of the quality of the movement. This information is still presented today, but engraved on the movement and visible only from the back of the watch, if at all.

Specific features of the watch may also be mentioned on the dial. For instance, "chronometer" signifies that the movement successfully withstood a precision testing protocol. If the watch is also prepared to resist superior depths, then that number may also show up on the dial. Some consumers appreciate when the presence of a major complication is indicated on the dial if not visible. This is the reason why some Patek Philippe models have the word "Tourbillon" on them.

Fig. 19: *Roman numerals*

Roman numerals

The antique Roman way of writing numerals still finds use today, especially in watches. An "alternative" system with its own rules, a certain question pops up very often about the way the "4" is written. In fact, this number can be written in two ways: "IIII" or "IV," and both are correct. Some experts say that the choice of the "IIII" depends on taste and legibility in order to balance out the "VIII" on the opposite side, also

Fig. 20: *Arabic numerals*

composed of four characters. It is easy to conclude that this solution allows the same quantity of symbols on each side of the dial if you divide it exactly in the middle.

Roman numerals were the first font used on timekeepers: you can still spot them on old church clocks. Today Roman numerals on watch dials reference a traditional, historical style.

Arabic numerals

Invented by Indian mathematicians, Arabic numerals progressively replaced Roman numerals during the Middle Ages with the advent of modern science throughout Europe. Today they are the standard, finding their way onto many watches.

Logos

Logos might be the most significant mark on a dial because they visually incarnate the brand. Consequently, logos give each watch an identity and origin. Such signatures are also one of the last indications to be removed when designing a "pure" dial.

Pay attention to logos. Most of them are the result of an important design process. Details, shape, and volumes (if three-dimensional) have all been deeply analyzed to be the most distinctive and best reflection of the identity of the brand.

Logos also express the personality of the brand. If one is famous for being dedicated to sports and performance, you should be able to see that, or at least feel it, through the logo. Other logos carry a message or embody a statement. Do you recognize, for instance, the Swiss brand with a crown in its logo? Metaphorically and thanks to this logo representing Rolex, everyone can somehow become the king or queen of his or her own imaginary kingdom. The Longines logo combines an hourglass with wings: the hourglass symbolizes timekeeping, while the wings may refer to the ability of the brand to measure short periods of time as Longines was one of the pioneers in competition chronometry. Longines has also developed watches boasting various inventions designed to aid pilots. Longines is well known for the "Weems", a flyback chronograph, and later "Lindbergh" models.

Apertures
Dials can be cut out to allow the displays of indications showing from underneath such as the date. These holes are called apertures and work like little windows specifically positioned to be legible and at the same time in harmony with the general design of the dial. About fifty years ago, they were much smaller, but they have gained in size over the last few years.

Running hands
If you pay very close attention to the way the second and minute hands move on a mechanical watch, you might be able to see that they progress in very small steps. This is directly due to the work of the escapement and its role in transforming linear energy into beats to measure time. While we know that the hands make little steps, the general impression is that a hand moves continuously. Such hands are called "continuously running hands."

Jumping hands
A "jumping" hand is different because it makes its steps from one marker to the next in a solid, visible jump. In other words, the hand (or disk for that matter) doesn't move smoothly but literally jumps from one position to the next. On mechanical watches in particular this might seem like just another way of making a hand move, but it is much more complicated to conceive and make than the observer might guess. The difficulty explains why a distinction exists here: Jumping systems are the result of a specific technical specialty that is really interesting to dig into. However, at this stage, just knowing that the two kinds of displays exist and that they are the result of technical phenomena inside the movement is sufficient. Most quartz watches combine both kinds with running hands for the hours and minutes and jumping seconds. This is due to the conception of the quartz movement.

Retrograde

A retrograde display means that the hand doesn't revolve around its own axis in the conventional fashion, but rather moves back and forth across an arc. When it reaches the end of the arc, the hand jumps immediately back to the start to begin the measurement all over again.

Sometimes, only one hand on the dial is retrograde, although there can be more. This one retrograde indication is enough to say a watch "comes with a retrograde display." Retrograde systems are often used to display the date, the month, or the power reserve. As with jumping hours, a retrograde indicator is a special technical construction. Even if it has become more common recently, it remains a type of display that adds another degree of fascination to the dial.

Three-dimensionality

For years and years, the unwritten rule among watchmakers has been to build the smallest and thinnest movements in order to make watches as discreet and wearable as possible. Paradoxically, in recent years, larger sizes for movements and watches have evolved so that this is no longer the norm. This new freedom has allowed watchmakers to express themselves in the design of movements and the introduction of three-dimensional displays. The contemporary brand Urwerk has built its whole catalogue around a three-dimensional hour and minute display using mobile cones or cubes and at times retractable telescoping hands.

Rare and innovative displays

Some creative and courageous watchmakers have also decided to go much further afield with unexpected and sometimes revolutionary displays. Franck Muller is one of them. In 2003, he invented the "Crazy Hours" watch. Why is it special? The numbers on the dial are positioned illogically. However, the hour hand jumps to the correct position every hour, indicating the true hour, while the minute hand works like a normal one. The effect is stunning: it looks like time moves backward and then accelerates.

In 2012, HYT launched the first watch able to display time using two unmixable liquids, one colored and the other transparent, moving inside a circular glass micro tube. This is an extreme challenge in terms of technology. The innovation is considered a major one in watchmaking at the start of the twenty-first century. Other watchmakers have decided to use magnetism to display the time in their creations: Christophe Claret launched, for instance, a surprising model in this vein in 2012. Named "X-TREM 1," the large rectangular watch bears two linear scales on each side. The left stands for the hours, the right for the minutes. The time is displayed by two metal balls moving inside sapphire crystal tubes that position themselves at the proper place along the scales. It looks like a magic trick, but the spheres are driven by magnets positioned inside the movement that moving along a rail. The weirdest thing is that if you shake the watch, the balls might fall to the bottom of the tubes momentarily, but they shift back to their correct positions thanks to the magnetic attraction.

Hands

Hands give life to a watch. Only by observing them can you distinguish if a movement is quartz or mechanical. And if you are really a connoisseur, you may even be able to deduce the frequency of the movement. Hands are strong visual markers of a brand, a watch, and even a collection.

Almost every kind of shape exists, and these can be painted, coated, skeletonized, engraved, and even gem-set. Their length is also variable. Some more spectacular hands are extendable and retractable; the Parmigiani Fleurier "Oval Pantographe" is a watch featuring such innovative and highly technical hands.

Hand shapes

Hands come in many shapes and sizes. A hand is normally straight, but many other forms exist. The serpentine shape comprises small, regular waves and was invented to display calendar information on a dial to distinguish this function from the hands dedicated to the main time display. This initial distinction has been forgotten by some brands, though, as in this day and age it is not unusual for aesthetics to come first.

Fig. 21: *Stick*

Fig. 22: *Leaf*

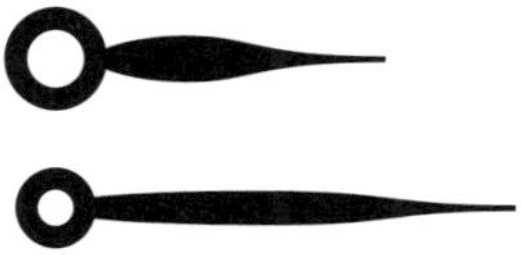

Fig. 23: *Sword*

Fig. 24: *Dauphine*

Fig. 25: *Breguet*

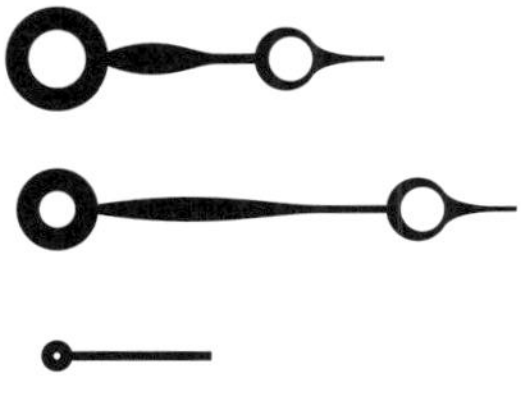

Fig. 26: *Mercedes*

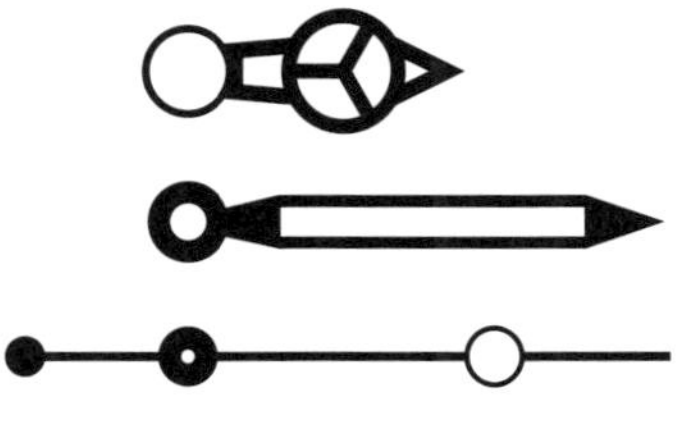

Anatomy
Display
Indications
Reading indications
Display modes
Hands
Hand shapes
Features
Enlightened displays
Functions and complications
Case
Straps and bracelets
Materials
Movement

Features

The length of the hands is also important primarily for visual harmony. Ideally, the hour hand should almost reach the indexes. The minute hand should extend all the way up to through the railroad track, and the second hand should reach the far side of the track. Color matters, too. Indeed, different colors are often used to distinguish functions.

Enlightened displays

Yes, you can read the time in the dark, but this is not a given on every watch. This ability must be added to the watch using particular materials. Electronic-digital watches are usually equipped with an automatic lighting system that can be activated by pushing a button. On analog watches, watchmakers had to be more imaginative.

They reached their goal by using luminescent materials on the hands and on the dial. This material, which lights up from a mix of solar energy and resin, and the most famous of which is the Swiss Super-LumiNova®, has the ability to store up and later emit the energy present in any kind of light. Depending on the quality of the material and the amount of layers applied, the result can be really amazing and strong. But it doesn't work like a lightbulb, and after a few hours the effect slowly fades away.

Another option is to use a special kind of gas sealed in micro tubes. In this case, the glow is constant because of a different physical effect. Tritium gas is inserted into a tube coated with luminescent material. The light comes from the interaction of the gas with the coating. Using such tubes implies more complicated work on these parts of the dial: The micro cylinders must be mounted on the hands or the indexes. Only a few brands like Ball Watch and Luminox, for instance, use them.

Functions and complications

Basically, a watch displays the hours and minutes at the very least. But it can do much more. Here we will explore the most common additional functions and complications available today. Everything that gets measured by a watch can be called a function. But when you investigate the world of watches, you also will hear a great deal about "complications."

Specialists have intense discussions about the definition of the word "complication." By default, a complication is an indication added to the hours and minutes and linked with time. Even the indication of sweep seconds can be fundamentally described as a complication as it requires adding some parts to the movement.

There are also complex mechanisms not linked with time measurement. Tourbillons and power reserve indicators are two examples. In these cases, the term "complication" doesn't apply and should be traded in for "indication" or "complex mechanism."

Basic functions

"Basic function" is a very common expression to describe particularly simple and common indications or complications. These functions also utilize a limited number of parts in a mechanical movement.

Date, day, and month

One of the most common complications is the date. It is usually displayed through an aperture on the dial. The numbers are printed on a ring that revolves underneath the dial, or the date can also be displayed with a hand.

Some watches additionally have a day indication and others months. The technical principle is the same, and all indications slowly switch to the next unit around midnight.

Big date

A "big date" is a more complex type of date. Instead of having a disk showing each numeral from 1 to 31, the date is composed of two disks, one each for the double and single digits. The display is powered by a more complex mechanical system.

And, often, big dates are highlighted within the dial design. For instance, a special frame can surround each number. Most systems work with two different disks. The disk dedicated to the tens is printed with numerals from a blank zone to 3, and the disk for the single digits from 0 to 9. Others work with small panels that hide one of the two disks when needed.

Moon phase

Almost all religions and cultures have used lunar cycles to determine important celebrations such as Easter, Chinese New Year, and the beginning and end of Ramadan. Today, moon phase indications are an aesthetic choice rather than a practical one. Usually moon phases are nicely decorated, sometimes with a human face for the moon. Some can be turned into pieces of art. They are often the only purely decorative complication, in contrast to the clear utility of the others.

A moon cycle lasts 29 days, 12 hours, 44 minutes, and 2.8 seconds. The most common moon phase indications are built with a disk making one revolution in two moon cycles. They are relatively precise as they are calculated on cycles of 29.5 days. Others can be more technical and exceedingly precise. In these cases, the cycles remain exact for 122 and even 1,024 years! Independent watchmaker Andreas Strehler has even conceived a moon phase that deviates by one full day in two million years.

In its most common and simplest version, the moon phase indication involves a disk bearing the depiction of two moons placed opposite each other. As the disk revolves, one moon appears as the waxing moon; when it is in the middle and fully visible, it depicts the full moon; as it gradually disappears it represents waning moon.

Anatomy

After years of remaining flat in order to fit into movements, the moon has recently become a sphere again in watches. Three-dimensional moon phase indicators have been introduced thanks to advances in machining. Though not the first to add it to a watch, De Bethune is a pioneer of the spherical moon phase display, and this indication has become a clear technical hallmark of the brand.

The moon phase moves very slowly in contrast with other complications. In addition, it is directly related with the night and therefore the world of dreams and magic. The moon is the celestial satellite most visible to the naked eye. This might explain why the moon phase is considered a particularly artistic or inspiring complication and a great platform for creativity.

Traditionally, the window allowing the view of the moon usually has the shape of a half-disk with two little bumps at the bottom, making it look like a moustache. Another system sees a little "stick" with a micro disk at one end that passes above a moon fixed on the dial, like on various Jaquet Droz models, for instance. Sometimes, a hand can also point to the different phases of the moon designed on a dial.

Nowadays, the most precious moon phase displays are made of gold or platinum and engraved or set with diamonds. Other disks display a face because it is said that if you look carefully at a full moon, you can see two eyes, a nose, and a mouth—the so-called man in the moon. Stepan Sarpaneva, an independent Helsinki-based watchmaker, has designed a very particular moon face for his timepieces based on his own visage, which is a strong signature element of his branding.

Day & night indication

Don't confuse a day/night complication with the moon phase display even if they can appear similar. Usually coming with an additional time zone function, the day/night complication is generally a disk that shows whether the reference hours shown in the time display are day or night hours by dividing it into two sections, one light-colored and sometimes containing a sun and the other darker with an optional moon. The disk makes one revolution in 24 hours. In fact, this is another ornamental complication, except in one case where the disk also indicates hours like on the "Rotonde de Cartier Day and Night." Naturally, day/night complications are reserved for watches displaying time on a 12-hour scale—which is the vast majority of watches.

Additional time zones

Watches can provide the time for different places on earth at the same moment. These systems have various names and can present different displays or functionalities, but the basic mission is the same: to indicate various times on a single dial.

Back in the early days of timekeeping, each city or region had its own local time. This is logical because time depends directly on the position of the sun, which differs according to the position on the globe. Still, the incredible number of local times led to a lot of confusion, in particular with advances in transportation methods; think about trains, for instance. Conflicting schedules and even accidents became frequent problems. Such issues led directly to global time measurement being divided into 24 time zones, which took place in 1884, the year of the Meridian International Conference in Washington, DC.

Each time zone corresponds to 15 degrees of longitude on the globe. Each zone shares the same local time from one end to the other. The worldwide reference point, called the Prime Meridian, is located in Greenwich, a district of London where the famous British Royal Observatory is based. This global system of timekeeping is abbreviated "GMT" in reference to Greenwich Mean Time. Local time is officially expressed according to Greenwich Mean Time as the reference (GMT +0). For instance, New York City operates at GMT -5, Paris at GMT +1, Singapore at GMT +8, Tokyo at GMT +9, Quito at GMT -5, and Hawaii at GMT -10. That said, some regions still maintain their own time, as time zones are as much a political decision as a geographical one. Such is the case with Venezuela at GMT -4:30,

Nepal at GMT +5:45, and since August 15, 2015 North Korea at GMT +8:30.

It was also decided to establish the switching point from one day to the next—a point called the International Date Line—at precisely 180° opposite the Prime Meridian. Also, called the antemeridian, this 180th meridian is the place where the current day ends and the next one begins. The imaginary line runs through the Pacific Ocean passing places such as the Fiji Islands. For those who remember the storyline to "Around the World in Eighty Days" by Jules Verne, you can now see exactly why the exploratory Phileas Fogg won his bet. He traveled from west to east, and because of that he "won" one day.

In 1960 the GMT system was adjusted. Basically, this reform concerned the way of coordinating the timing system in place, taking solar time into account to manage small gaps that occur each year, which are compensated for by the introduction of leap seconds. The current usage is known as "UTC," which stands for Coordinated Universal Time. Nevertheless, the GMT anagram is still common in watches that display at least two time zones.

GMT and second time zone

Watchmakers have invented many different ways to express time in various places around the world simultaneously! The most common system, still called "GMT" or "second time zone," involves at a minimum a second display for the hours—usually as a hand making one revolution in 24 hours. In other words, instead of a first loop from midnight to midday and a second one from midday to midnight, midnight remains at the 12 o'clock position but midday is now at 6 o'clock on the dial. The minute hand naturally remains correct. Usually, a scale or a bezel helps in reading the second time zone. Other kinds of displays are also in use, like that of the revolving disk on the Rolex "Sky-Dweller."

Some interesting models include pushers that can make this 24-hour hand go forward or backward in a very comfortable way.

GMT/second time zone watches offer usually only one additional hour display. This is useful for the owner who needs to know the time at home and the time where he or she is traveling. Therefore, those times are called "home time" and "local time." But other GMT models offer multiple additional time zones.

Worldtimer

The oldest system for displaying multiple time zones is called a "worldtimer." It works using two concentric rings on the edge of the dial. The outside one, divided into 24 sections, is stationary. Each corresponds to a time zone generally identified by the name of the main city in that zone: London for GMT +0, Paris for GMT +1, Los Angeles for GMT -8, etc. The second ring is positioned inside the first and performs a full revolution in 24 hours. It is also divided into 24 equivalent zones and numbered from 1 to 24 to represent the hours of the day. To make it easier to read, watch-makers generally give its background a light color between 6:00 and 18:00 and a dark color between 18:00 and 6:00 to distinguish times of day and night. Sometimes, brands go further and add a little sun or moon instead of the "12" or the "24" to avoid any confusion.

This display has the advantage of offering up the time in any region at one glance. However, with this kind of display, you need to pay special attention when you want to check the precise time somewhere on earth because of the quantity of data: the scale is quite small and might even show Daylight Savings Time!

Anatomy
Display
Indications
Functions and complications
Basic functions
Date, day, and month
Moon phase
Day & night indication
Additional time zones
Power reserve indicator
Rotating bezel
Stop seconds
"Regulator" display
(Highly) complicated
Case
Straps and bracelets
Materials
Movement

Power reserve indicator

Nobody likes to discover that his or her watch has stopped because it ran out of power. In addition to not knowing the current time anymore, you need to set the watch again. This can be really tedious, depending on which complications you have to set, too. For this reason watchmakers invented the power reserve indicator, which works just like your car's petrol gauge because it tells you how much energy is still available in the movement. Many different kinds of displays exist, the most common being a little hand on the dial.

Rotating bezel

The flat raised ring surrounding the dial is called the bezel, and it can be stationary or revolving. A rotating bezel shouldn't be considered a complication as it is not included in or driven by the movement. Nevertheless, it can add functionality to the watch.

Divers need to know how long they can, or must, stay at a certain depth to make a secure dive and avoid decompression troubles. In this case, the diver can use the rotating bezel to check the duration of time spent under the water.

Moreover, according to the ISO 6425 norm concerning diver's watches, this is a compulsory element on every watch called a diver's watch. Consequently, on diver's watches, the rotating bezel must move notch-by-notch and counterclockwise only.

The starting point must also be clearly marked and visible in the dark. In addition, the first 15 minutes must be individually marked, after which five-minute intervals are sufficient. Finally, it has to be movable with the palm of the hand as divers usually wear gloves.

Rotating bezels can also have other purposes. For instance, they can be part of the command system to set complications. On the "Yacht-Master II regatta" watch by Rolex, the bezel runs the mechanical countdown system; turning the bezel allows the user to set the countdown mechanism. Note that command bezels are very rare and involve a high level of technical finesse.

Stop seconds

Watches with a stop-second system can be perfectly synchronized with the current time as this complication actually stops the second hand when the crown is pulled. It is a useful function as standard movements don't work that way. When you want to set the time on an ordinary watch, you pull the crown, which allows you to move the hour and minute hands. But the second hand continues to run. With a stop-second system, the second hand pauses, too. Many purists like this system, which is also known as hacking seconds, because it makes the precision of their watches easier to verify.

"Regulator" display

The "regulator" display harks back to the main clock found in a workshop that provided the reference time to which all watches in production were set and adjusted. To ensure its precision and avoid any confusion while reading the time, the position of the hands corresponded to the location of their wheels inside the movement. Thus, the hands were not coaxial—meaning on the same axis—as is usually the case, but dispatched to different sections of the dial. The main configuration of a regulator display today comprises presenting the minute hand in the middle and the hour and seconds hand in distinct subdials.

(Highly) complicated

Some functions are much more complex and hard to build and set than others are. One basic measure of a watch's level of complication is the quantity of additional components integrated into the movement. A high complication needs dozens, and even hundreds of them! For instance, a movement with a chronograph needs about 250 more components.

Remember that each part represents a lot of time for conception, manufacture, assembly, quality control, and adjustment. Each part also represents a potential problem and many unnecessary costs if poorly manufactured. Consequently, a complicated watch is much more expensive and takes longer to produce than a simpler movement. The size of the company in terms of production facilities also has an impact. Big brands are used to industrializing a maximum of tasks and dealing with large-scale volumes and costs. But smaller brands can't or don't have the same means at their disposal. So, if you consider that new movements must always be tested for a long time before being considered reliable and that the cost of production decreases a lot depending on the volume, you understand why smaller brands can't be as financially accessible as you might like them to be.

Number of parts, costs, size of the firm, and innovation are all elements to bear in mind when considering complicated watches.

Basically, major complications could be divided into five main families:

- improvement of precision;
- measurement of short periods of time;
- calendars and astronomical displays;
- striking watches;
- and contemporary ultracomplex mechanical systems.

Chronograph

The chronograph is the best known and most widely produced major complication. It is indeed so common that we often forget that it is very difficult to create, build, and adjust. A chronograph is designed to measure short periods of time on demand—like laps in a race, for instance. This means that devices—usually pushers—are used to start, stop, and reset the timing mechanism without stopping the main movement. Sometimes, a single pusher can be employed for all three actions; then the watch is called a monopusher chronograph—the oldest type of chronograph. Today, this configuration is less common as the majority comprises multipusher systems.

For optimal legibility, the chronograph's second hand is the longest and positioned in the middle of dial. The chronograph's counted minutes and hours are usually displayed on smaller dials called counters or totalizers. This is the most frequent configuration, but many variations do exist. When you get your hands on a chronograph for the first time, the goal should always be to locate which indications are dedicated to the main time and which are devoted to the chronograph.

A chronograph system works like a timer by feeding off of its own second, minute, and sometimes hour counters. It pumps its energy from the caliber. However, the chronograph is one of the few

Anatomy
Display
Indications
Functions and complications
Basic functions
(Highly) complicated
Chronograph
Calendars and astronomical displays
Equation of time
Chiming watches
Striking watches
Musical watches
Alarm
Tourbillon
Astronomical complications
Depth gauge
Case
Straps and bracelets
Materials
Movement

complications that is not continuously tied to the movement. Every time you start the chronograph, you connect it to the main caliber to pull the necessary energy. Of course, depending on the system used, activating the timer can have a small influence on the movement's precision by slightly slowing it momentarily.

A chronograph needs a coupling system, just like a clutch in a car, to master connecting to and disconnecting from the movement. Three kinds of systems can be used: the first one is the so-called vertical clutch, the second one is horizontal, and third works with an oscillating pinion. These names originate in their respective structures and are activated by components that make the link between the pushers and the coupling systems. And again, two versions exist: the simpler ones use a cam while the more technical, reliable, and precise chronograph utilizes a column wheel. This three-dimensional wheel activates or disengages levers by revolving.

Of course, all kinds of displays can be used for chronographs: hands are the most common, but digital indications exist, too.

Flyback chronograph
A flyback chronograph is a special type of chronograph that allows the reset button to be used during a measurement to stop and automatically restart the chronograph.

This means that the wearer only has to push the button once to immediately start a new measurement instead of first stopping the function, resetting it, and starting it again. The flyback function is useful when you need to move as quickly as possible from one measurement to the next. It was developed at the beginning of the twentieth century especially for pilots, who needed this functionality for aerial maneuvers.

Split-seconds chronograph

The most complicated timer is the split-seconds chronograph, also called a "rattrapante" (this name comes from French). This type of chronograph allows the measurement of different intervals of time within the same measurement. Two events starting at the same time, but not lasting the same amount of time, can both be tracked. For instance, you could separately measure the timing of two sprinters and also find out the difference in time between them. The mechanism allows this because it combines two superimposed second hands. When the chronograph is started, both begin to move synchronously. When the start/stop button is pushed the first time, it will stop the first split-seconds hand, but the second one will continue moving. Another press of the start-stop button will stop the second hand. You can then read the two measurements and the difference in time between them. Pressing the reset button makes the two hands return to their initial positions at 12 o'clock. The mechanism is then ready for a new time measurement. The split-seconds chronograph can also take various interval measurements during the same session. Indeed, the second hand can rejoin the main second hand during a measurement to determine a new time interval by pressing the reset button. This can be repeated again and again.

Some brands go further and include more than one split hand. The A. Lange & Söhne "Double Split," for example, boasts two and can even measure "split" minutes. Extremely complex to conceive and build, such chronographs are considered very rare high-end timepieces.

Calendars and astronomical displays

Every indication linked to the date can be considered a calendar complication. We have already discussed day and date complications, which are the simplest ones because, technically, they integrate only a few components. But in this case, simplicity has an impact in terms of use. Day and date must be set at the end of every month that doesn't last 31 days. Watchmakers have invented complex mechanisms in order avoid all that work: so-called annual and perpetual calendars must be set respectively only once a year or only on March 1, 2100!

Annual calendar

An annual calendar automatically adjusts itself at the end of months comprising 30 and 31 days. This means that you have to manually adjust it only once a year at the end of February, to account for February's 28 or 29 days. Before this complication appeared in the 1980s, consumers only had the option of choosing between a standard calendar that must be adjusted five times a year or the very complex and expensive perpetual calendar.

Nowadays, annual calendars show up in many brands' catalogues. More practical and technical, they are a good in-between choice. A competition of sorts has also emerged among brands and watchmakers to create the simplest annual calendar mechanism. Watchmaker and scientist

Ludwig Oechslin invented one that adds only four components to the movement. The "Sky-Dweller" by Rolex, which came later, also only adds four components to the movement for its calendar.

Perpetual calendar

A perpetual calendar automatically adjusts to the difference between 30- and 31-day months and also takes into account the longer February of leap years as well as the 28-day February of the normal cycle. This mechanism actually integrates a mechanical memory—a part called a cam—which reproduces the four-year cycles that include leap years. Nevertheless, the majority of perpetual calendars on the market will have to be adjusted again on March 1, 2100. This is because 2100 will be a secular year instead of a leap year—the secular year, which doesn't leap, occurs every time the first year of a century is not divisible by 400.

Perpetual calendars belong to the most prestigious major complications among connoisseurs. They are easy to recognize thanks to a leap year display on the watch's dial or case back. It is a very finicky function to set even today. However, new versions have been designed for more comfort, with setting managed directly by the crown or through use of simple correctors and without risk of damaging the movement.

Equation of time

The time we see displayed on our watches is based upon what is called mean time. Because—surprise!—the time shown there is actually correct in a solar sense only four times a year, around April 15, June 14, September 1, and December 25. On these dates—which are the equinoxes and the solstices—the time indicated by the sun precisely corresponds to the mean time on a watch, which is an average time calculated by making equal divisions of hours, minutes, seconds, and so on. Why does this variation exist? It is due to the fact that the earth revolves around an inclined axis—there would be no seasons without that—and also because the earth's speed of rotation around the sun varies thanks to the planet's elliptical path. The combination of these astrophysical factors creates a variation that makes days sometimes shorter or longer as compared to mean time, also called civil or standard time. Mean time is an average based on dividing 24 hours into equal durations.

The equation of time complication displays how many minutes must be added or deducted from mean time to get the actual solar time. The scale goes from minus 16 minutes and 23 seconds to plus 14 minutes and 15 seconds and depends on the location of the watch on the globe.

The equation of time has no real practical use in itself and is mostly deployed in perpetual calendars as a supplementary and coherent indication. Brands must offer customization of the equation of time depending on the owner's exact position on earth.

Chiming watches

Some watches also "play" music: in addition to displaying the time, they can express it with sound. This sound can be melodies or chimes produced by gongs or bells. Complex, even very complex, most of these present great challenges in manufacture and assembly. The movement and the case require special attention because they work closely together to produce the quality of the tones. Like a violin, the case amplifies the sounds, making them as audible as possible. Consequently, in addition to marking time, these watches must also be tuned to create correct, precise, and pleasing sounds and notes.

Striking watches

If you have ever lived next to a church, you may have learned how to interpret how the bells chime to tell time. Striking watches express time through tones following the same principles.

The type known as a "minute repeater" outlined next is the most common, and precise, one, but a variety of versions exist. For instance, some models can ring to the precision of five-minute periods, others only on the quarter hour, or even just the hours.

How to "listen" to the time
The minute repeater system involves two different tones: one low, the other high. The low ones are played first to chime the hours (dong). Once that finishes, the quarter hours are indicated by a double tone: low and high in succession (dong-ding). The high tone then chimes the minutes (ding) after the last quarter hour has finished. So 11:56 will sound like this:

- 11 low (dongs) for 11 hours;
- 3 low-high (dong-dings) for three quarter hours (45 minutes);
- 11 high (dings) for the 11 last minutes (from :45 to :56).

This way of "telling" time is the most common in acoustic watches.

Striking mechanisms
A chiming mechanism requires energy to run either "on demand" or automatically at a determined moment. An independent spring barrel is powered either by hand using a lever that slides along the case or by the movement itself. Then, sensors mechanically read the time on cams connected to the movement. They activate the striking system in succession for the different indications: hours, quarter hours, and minutes. To obtain a sound, a little hammer knocks on what is called a gong, which is a wire made out of solid metal and which usually has a circular shape to fit around the circumference of the watch case's interior. Its length determines the depth of the tone. In some case, it is tuned manually by watchmakers to obtain just the right notes. Finally, a regulator controls the flow of energy to ensure that the rhythm of chimes and the tonalities are consistent from beginning to end.

Striking versus repeating
Both the terms "striking" and "repeating" convey the same category of watch in the sense that they indicate time shared by sound. The difference is that a striking watch activates automatically in contrast to a repeater, which sounds the time only on demand. Often both systems are combined, which provides a striking watch with repeater functionality.

Minute, five-minute, and quarter repeater
A repeater watch sounds the time on demand. Consequently, the wearer must activate the mechanism by pushing a button or pulling a lever. This action winds the mechanism to generate the necessary energy. Then, the system starts working automatically.

There are various kinds of repeaters. These differ according to the interval of time they have been designed to chime. For instance, basic ones chime only on the quarter hours; others sound periods of ten, 7.5, or five minutes. The most refined watches are able to chime hours, quarter hours or ten-minute periods, and minutes. These are called minute repeaters.

Anatomy

Small strike and grand strike
Small strike (called the "petite sonnerie" in French) refers to a watch that systematically rings hours and quarter hours "au passage" ("in passing"), which means every time the minute hand arrives at the determined hour. A "grand strike" ("grande sonnerie") watch also rings the hours, but repeats the current hour before chiming each quarter. Some systems also contain an on-demand system like the repeater. Most of these models also integrate a silence mode as well.

The grand strike is a very prestigious complication because of the myriad of technical constraints it implies. Just imagine that it has the ability to chime no fewer than 35,040 times per year! Very few brands or watch-makers are actually able to build such a complication. And experts agree that it is the Mount Everest of the classic art of watchmaking.

Grand chime (Westminster) and cathedral gong
On a grand chime striking watch, the quarter hours are played on at least three, but mostly four, or even five, gongs instead of two like the usual striking watches. Each quarter hour has its own melody using the notes C, D, E, and G. The melody that is the most famous and frequent is called the Westminster chime because the bells in the Elizabeth tower of the Westminster Palace play it. The word "cathedral" can be applied to a gong and signifies one that is much longer

than usual. Generally, a gong makes more or less one loop around the movement. For a cathedral chime, it is longer and can reach two loops. The superior length creates a better quality sound that resonates more. These are frequently used with chiming watches.

Musical watches

Musical watches are similar to music boxes and integrate a mechanism able to play a full melody. Usually, the music can be scheduled to start automatically at a certain time or it can be activated on demand by pulling a lever or pushing a button. "The Stranger" by Ulysse Nardin integrates such a system, which is directly visible on the dial.

Alarm

Wake up! Some watches include an alarm—and you better know how to set it to avoid a rude awakening at the wrong time. This function is a simplified version of a striking mechanism, using only one hammer and gong. When it starts, the hammer knocks the gong continuously until the stock of energy stored comes to an end or it is stopped manually.

Most alarm watches can be set only 24 hours in advance. But there are exceptions, like the "Glashütte Original Senator Diary," which can be set up to 31 days in advance. The quality of the sound counts for less than its intensity here. In other words, efficiency replaces beauty.

Tourbillon

To stay precise, a mechanical watch movement must work as regularly as possible. This means that any perturbation has a negative impact on its chronometry. Among these perturbations even gravitational force plays a role. Its pull can influence the hairspring and the oscillations of the

Anatomy

Display
Indications
Functions and complications
Basic functions
(Highly) complicated
Chronograph
Calendars and astronomical displays
Equation of time
Chiming watches
Striking watches
Musical watches
Alarm
Tourbillon
Astronomical complications
Depth gauge
Case
Straps and bracelets
Materials
Movement

balance wheel. Instead of beating in a regular, concentric way, curves can deform, impacting accuracy. In addition, a wheel is never perfectly balanced across its entire circumference: there is always a heavier or lighter spot because of imperceptible variations in form and mass. As with car wheels, they must be balanced.

At the beginning of the nineteenth century, a Swiss watchmaker established in Paris named Abraham-Louis Breguet had an idea that would help control these issues. His concept was to average out the potential deviations and errors by making the whole regulation system revolve around one central axis. In that way, accelerations, slowdowns, errors, variations, and perturbations would be all mixed together to produce an average rate. And this average rate would be more stable, consequently making the watch more precise. So Breguet developed a system with a revolving structure called a cage to hold the escapement, balance and hairspring (together known as the regulating system) inside. To sum up, a tourbillon designates a movement in which the regulating system is not fixed but revolves around its own axis.

This element was particularly clever because at the time of its invention all watches were pocket watches kept in the same vertical position almost all the time. Because of that, the movement was particularly exposed to gravitational force, making the tourbillon particularly useful.

That said, the tourbillon has been proven to be useful on wristwatches, too, despite the fact that they continuously change position. Brands like Greubel Forsey made a specialty out of the tourbillon, with results matching their efforts.

Today, although removed from the quest for precision, tourbillons are very much appreciated and desired. And there is a special reason for that: the rotations of a tourbillon are so fascinating that it makes watches seem almost alive. And this has a magic power of attraction.

In addition, the tourbillon used to be a very difficult construction to build, assemble, and adjust. Today, though easier to make using modern processes, it still symbolizes a certain level of prestige and know-how. Rest reassured that many brands continue to constantly innovate and improve upon the tourbillon with ever more complex and spectacular constructions.

Not a complication, but complex (!)
A tourbillon is sometimes mistakenly labeled a complication. Why? Because of its complexity, the tourbillon is one of the most difficult technical constructions to realize in watchmaking—and the only one that is not really a complication: the tourbillon is a "complex" mechanism, and not a "complication." Even if machines and technologies make it much easier to produce in the modern era, its regal aura remains, and for good reason. Today some specialists (and particular brands) are keen to identify the tourbillon as a "functioning complication," meaning a technically complex mechanism dedicated to improving the rate of the movement. They similarly include the chain-and-fusée system in this category, a construction using a micro chain to regulate the force provided by the barrel; it helps the escapement maintain a very regular flow of energy.

Flying tourbillon
The tourbillon is usually positioned between the base plate on the back of the movement and another removable part called a bridge. On a flying tourbillon, the cage is secured on the back only. The upper bridge is taken out, though it is more securely attached and strengthened on the one side. This structure has the advantage of providing a better view of the mechanism inside its cage to the wearer. It also allows the watchmaker to position the tourbillon slightly above the surface of the movement for better visibility.

Multiple tourbillons
If a single tourbillon helps with chronometry, what happens when several tourbillons are placed within one movement? Some watches contain several tourbillons, which they do in order to impress and fascinate the consumer. These tourbillons are usually interconnected with the objective of averaging out their respective rates.

Anatomy

Display
Indications
Functions and complications
Basic functions
(Highly) complicated
Chronograph
Calendars and astronomical displays
Equation of time
Chiming watches
Striking watches
Musical watches
Alarm
Tourbillon
Astronomical complications
Depth gauge
Case
Straps and bracelets
Materials
Movement

A spectacular example of a multiple tourbillon is the “Tourbillon of Tourbillons” by Antoine Preziuso.

Multiaxis tourbillons

Sometimes the cage of the tourbillon itself can be integrated into various rotating systems. Thus, the escapement revolves not only along one axis, but two and up to three. These configurations aim to better mix the different positions of the regulator in order to increase precision. One of the most impressive is the third generation of the “Gyrotourbillon” from Jaeger-LeCoultre, which sees the tourbillon revolving in three dimensions. However, the cage is still only secured to the movement by a single connecting structure, which makes it a flying tourbillon, too.

Astronomical complications

Time measurement comes straight from astronomy—in others words, the observation of planetary motion. Today, some watches put the planets and night sky on their dials. Some representations are stationary; others are animated. The moon phase is the most well-known astronomical complication. Astronomical watches belong to another, much more exclusive and complicated category. The most sophisticated of these watches are able to reproduce, in different forms, the motion of stars and planets.

Night sky

Some dials, and even case backs, reproduce the night sky as it appears from a certain position on earth. Most of the time, the owner's location is chosen and these dials are works of art in terms of craftsmanship. They can move or stay fixed in one position.

Planetarium

A planetarium is an animated construction representing the motion of our solar system's planets. Usually, they are big mechanisms that are positioned on a table or other platform to be exhibited. But watchmakers have succeeded in miniaturizing planetariums to fit them into a wristwatch. One of the most spectacular of these is the "Planetarium" by Van Cleef & Arpels executed in conjunction with Christiaan van der Klaauw.

This watch represents our solar system with the sun in the middle and six planets revolving around it at their respective speeds. Saturn, for instance, takes twenty-nine years to make an entire revolution around the dial.

Depth gauge

A depth gauge is a diving mechanism that indicates how deep you are underwater. It is a very rare functionality in watchmaking because its precision is not reliable enough to be used as a safe instrument. Nevertheless, designing a depth gauge is a fantastic challenge for engineers, and some have been really cleverly conceived. A depth gauge is also not a watchmaking complication as it is not connected to the movement and has no link with the indication of time.

Case

The case is the shell of the watch that protects all of its sensitive parts—the movement, the hands, and the dial, in particular. The case has to be absolutely solid to resist shocks and brusque movements. It also has to be water-resistant to avoid incursions of liquids, dust, and dirt, which is a real challenge because a case comprises different parts that are very often manufactured independently. In addition to this first functional role, another fundamental mission of the case is to also be aesthetically pleasing and ergonomic. Finally, each case integrates at least one aperture for the setting system—chiefly a crown situated on the side. But some have systems where the time and functions can be set from the back or through another part like the bezel, for instance.

A case is therefore a complex arrangement of multiple components and elements depending on the integrated functions and design chosen. Rotating bezels are present on diver's watches because they are fundamental to using the watch underwater. Chronograph watches, for example, must have buttons for the start/stop function and resetting. Very complicated watches like mechanical perpetual calendars are outfitted with tiny recessed correctors to set and adjust the complication.

Case shape

The vast majority of watches sold have a round case shape. Indeed, round used to be the standard shape of any instrument with counters. Many options exist for case shapes today.

Round Fig. 27

Round watches are the most common on the market and remain the most popular. This is the very traditional shape of watches. Some experts also say that the round shape is due to the fact that watches used to be transportable instruments. In contrast, other instruments fit into boxes or cases looking like furniture with squared or rectangular shapes. However, the round shape appears to simply be the best solution to read time because it creates a harmonious casing for the movement and complements the circular motion of the hands.

Shaped watch

Every watch that is not round is what an expert would call a shaped watch. But these represent the absolute minority. Among them, the square and tonneau shapes are the most common, but every kind of form is possible. From the rectangle to the triangle, from the integration of various forms in one single final design, everything is an option. For instance, the "Crash collection" by Cartier can be considered a good example of what is possible in terms of creatively shaped cases.

Square Fig. 28

Somehow, the square case plays the role of the nonconformist, like a perfect opposite to the traditional round case. These cases are actually rare because they often look quite Bauhaus in their form, in which no curve is allowed. Not everyone likes it, but if you do, you will be absolutely satisfied with such a shape.

Tonneau Fig. 29

Tonneau, which means "barrel" in French, is a shape that has stood the test of time and still much appreciated. The shape really corresponds to that of an almond, the extremities of which have been cut off. Its difference from conventional watch shapes makes it really special, and you should make sure you are very attracted by such a watch before purchasing it. Whether you call it dandy, extravagant, unusual, or just different, the tonneau shape belongs in its own singular category that can be summed up as a freed classical style.

Cushion shape Fig. 30

The cushion shape is a very subtle mix between round and square. A round dial and a square case, with smoothly curved sides, characterize this form. As with the tonneau shape, the cushion shape is as rare as it is special.

Cuff watch

Usually, the bracelet plays a supporting role in securing the watch. On a cuff watch, this relationship works differently: The bracelet is as important as the watch itself. The strap becomes such a major element of the watch that it gives it an absolutely unique personality. Cuff watches are particularly prominent in the category of jewelry watches. Mostly large and made of precious materials, they also express the creativity and know-how of a brand, particularly in gem-set bracelets that can look like spectacular ribbons of diamonds and precious stones.

Cuff watches are not only a woman's privilege, however. On the contrary: When it comes to men, things get serious, but in another way. Large, muscular, solid, cuff watch straps are generally made of leather and composed of two layers. A first one, tight on the wrist, works like the bracelet while the second one secures the watch. Initially, this kind of construction was dedicated to demanding situations like exploration or flying. Today, it is, naturally, a question of style.

Fig. 27: *Round*

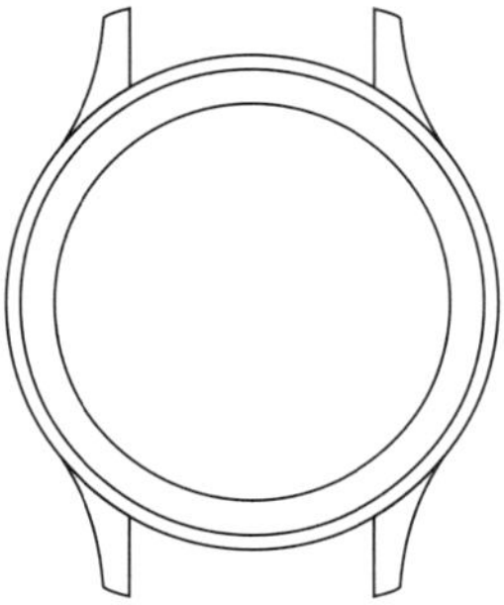

Fig. 28: *Square*

Fig. 29: *Tonneau*

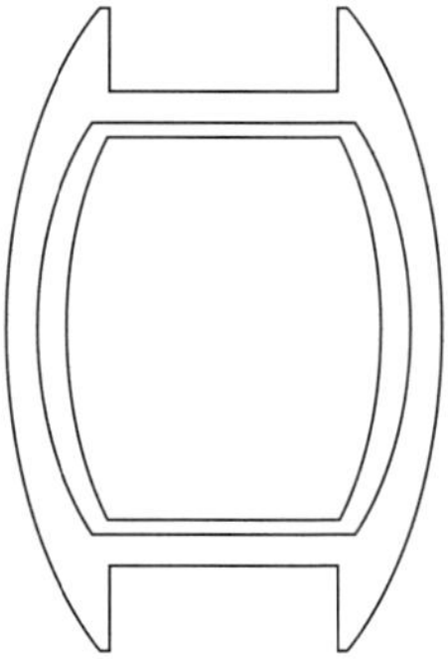

Fig. 30: *Cushion*

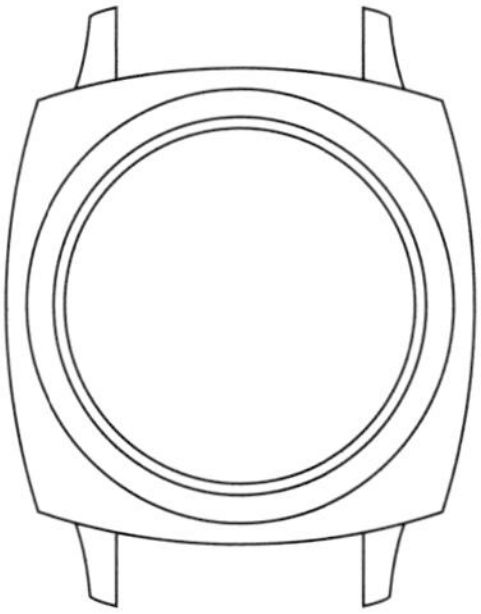

Fig. 31: *Creative shapes*

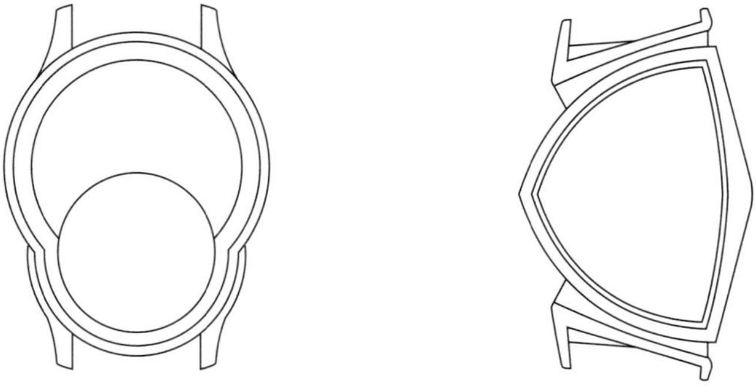

Anatomy
Display
Indications
Functions and complications
Case
Case shape
Bezel
Crown
Location of the crown
Screwed-down crown
Locked crown
Protected crown
Pushers
Alternative setting systems
Lugs
Case back
Indications and seals
Straps and bracelets
Materials
Movement

Fig. 32: *Bezel*

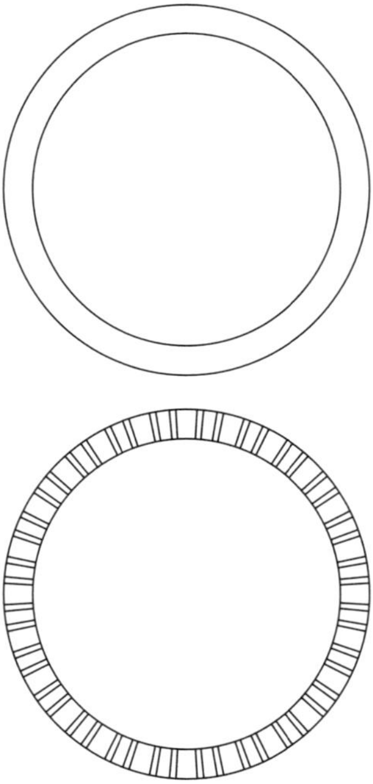

Bezel

The bezel is a ring that surrounds the crystal on top of the case. First and foremost, it helps keep the watch water-resistant. Secondly, it is also an element contributing to the look of the watch. It can be colored, decorated, engraved, or made of a different material than the rest of the watch. Aesthetically, the bezel gives visual structure to the watch. Bezels can display scales, names, and symbols. For instance, the legendary Omega "Speedmaster" has a tachymeter scale on its bezel.

And just think about the "Royal Oak" by Audemars Piguet. Launched in 1972 and revolutionary in terms of design, its bezel is octagonal and held in place by eight screws.

Finally, bezels can rotate. In this case, they become functional, meaning that they participate in timing measurements or access to functionalities.

Crown

The crown is the "remote control" of the watch. It is a small flat cylinder with a surface that allows a good grasp using knurling, geometrical engraving, or rubber. On the top, you usually find the logo of the brand, a decoration, or perhaps even a gemstone. The basic functions of the watch like the time and date can be set by the crown. On manual winding models, it also provides energy to the movement when the wearer turns it back and forth between the thumb

Fig. 33: *Crown*

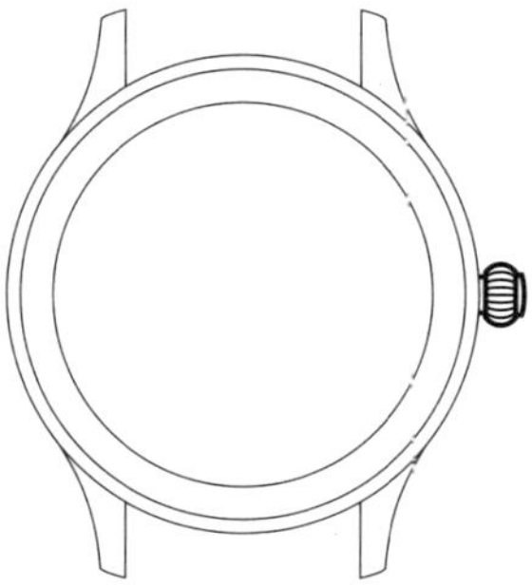

and index finger. The crown is a sensitive component that must be handled with a lot of care. For a long time, it has presented watchmakers with the challenge of preventing water or dust from entering the watch at the place where the crown stem extends from the case.

When not in use (what's usually referred to as position 1), the crown is positioned just next to the case. This is also the position that allows the wearer to wind the movement manually. When you want to modify settings, like set or adjust the time, you have to pull the crown out. Depending on the model and the quantity of functions, you pull it by notches; two notches are usually the maximum. Each notch corresponds to one or two functions. In that case, turning the crown in one way sets one function, and in the other way another.

Usually, a watch has one single crown. But it can have more depending on the functions of the movement.

Location of the crown

The crown is usually positioned at 3 o'clock. But a crown can be situated also at 12 or 9 o'clock. The 12 o'clock position is reminiscent of the first wristwatches, which were, in fact, pocket watches modified to become wristwatches with the addition of a set of lugs to secure the strap. The 9 o'clock position is more exotic and can be helpful when the left hand remains lifted up as while driving, for instance. The crown then doesn't hurt the wrist as if it were positioned on the right side of the case.

Screwed-down crown

A screwed-down crown is like a crown on steroids. It is indeed screwed in (and not just pushed) to avoid any impurity getting inside the case and ensures the watch is water-resistant. This is possible thanks to a fixation system that includes a threaded tube on the side of the case on which the crown gets screwed. Often, a spring pushes out the crown once open to facilitate manipulation.

Locked crown

Another approach to make watches resistant to external particles is the locking system. The principle is to push the crown strongly against the case to create a hermetic seal. The historical Italian brand Panerai developed a system with a lever that pushes and blocks the crown (in evidence on the Luminor collection). Jaeger-LeCoultre found an alternative in what it calls the compression key: this element situated between the case and the crown works like a lock. By turning the key 90 degrees, the crown is freed and you can use it as usual. To lock it, you have to push it against the case and turn the key back.

Protected crown

As you might guess, a crown is a vulnerable part of the watch. A shock can damage or even break it.

This is why—specifically on sporty models—there can be little spurs on each side of the crown to function as protection; these can be directly tooled into the case or added later.

Fig. 34: *Pushers*

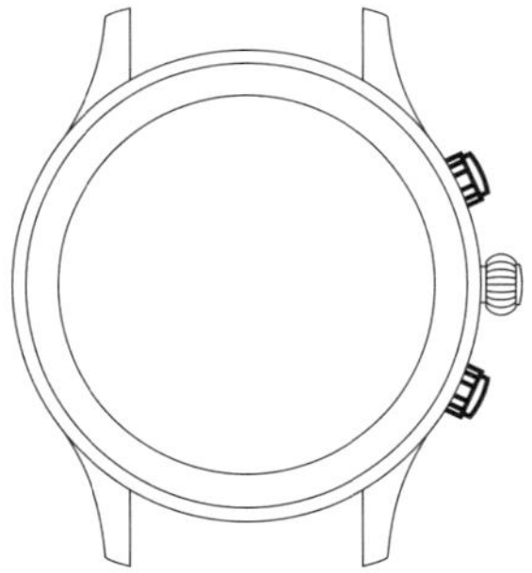

Pushers

Some watch functions require pushers in order to be used. The chronograph is the most common instance, and it generally works using two pushers: One above the crown and the other below.

A pusher is basically a button outfitted with a spring to make it come back to its initial outward position. It can be locked by a dedicated system, usually a screw ring or a cylinder that surrounds it.

Sometimes, pushers are integrated into the crown, as on single-pusher (monopusher) chronographs.
The previous traditional shape for pushers was the so-called "mushroom" shape: a round, tiny, and almost flat disk hanging over a vertical bar. Nowadays, the most common shape is the cylinder or the rectangular block. And remember: never use pushers under water as it may enter inside the watch.

Alternative setting systems

Sometimes, functions require dedicated setting systems—like that of the perpetual calendar. To avoid having too many buttons on the case, watchmakers invented correctors for use with the perpetual calendar. Correctors can be sensors or very small and recessed pushers that help set and adjust the specific functions. In the case of sensors, you need a little nonmetal stylus that usually comes with the watch to press them. Avoid using a ballpoint pen as the ink might color the case and eventually damage the internal gaskets. Blancpain conceived a smart system called the under-lug corrector. It has a micro lever coming discreetly out of the case, allowing one to set the function using a finger.

Fig. 35: *Lugs*

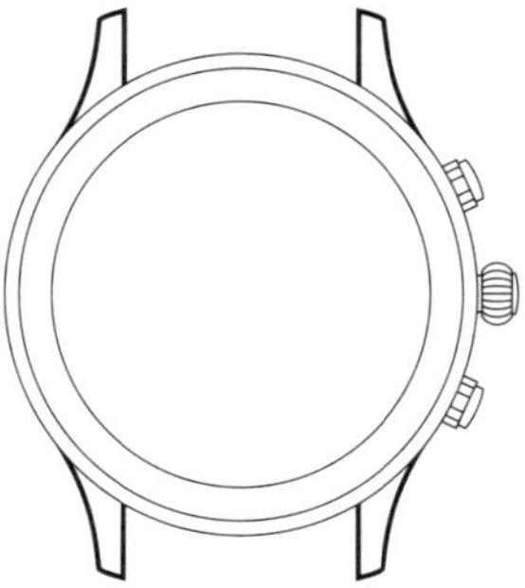

Lugs

Lugs are the official names of the case protrusions used for attaching a strap. They first appeared when watches began to be worn on the wrist instead of in the pocket. The first lugs were in fact solid wires of metal soldered onto a standard pocket watch case. The form changed rapidly to become two spur-like chunks of metal popping up from the case. This is still the main morphology used today. At the beginning, lugs were quite long and clearly extended beyond the case. But they quickly became a main element in watch design, their form, length, and curve becoming significant. On certain models lugs have also become a visual signature. The goal of a lug is to harmoniously connect the case to the strap and to maintain it securely on the wrist.

Flexible lugs

Usually lugs are firmly attached to the case. But brands have invented flexible systems that make the lugs move slightly to adapt better to the wrist and, consequently, offer more comfort.

Attachment systems

Plenty of systems exist to attach the lugs to the strap. The most common one is the spring bar that goes through the strap and penetrates tiny holes inside the lugs. An alternative system sees the bar become a micro tube screwed onto each lug.

Lugs can also seem to disappear from the case with certain designs. If it looks this way, it means that the case has been designed to have no gap between it and the strap. This type of design is called "integrated," which means that the strap is integrated into the case. It affords perfect continuity between the case and the strap or bracelet.

Case back

The back of a watch is the side opposite the dial. This side also contains the case closure, placed on the reverse side to protect the interior. The simplest case back is a single layer of the same material as the rest of the case; it is generally curved. If that material is metal, sometimes people choose to engrave a personalized message on it, as a watch is often a gift for a special occasion. Some are decorated with engravings, tributes, and also sometimes handcrafted works such as enamel. Nowadays, the trend is clearly to "open" the case back to reveal the movement beating inside. To do so, a transparent crystal glass—either made of mineral or sapphire crystal—is integrated into the case back. If it is "open," the case back comprises a ring or gasket that secures and frames the crystal disk and links it with the case.

On certain watches, the case back can comprise a hinged double layer so that the movement may only be seen on demand by opening it. The direct case back is usually transparent to allow a look at the inside of the watch. Covering this is a second one that hinges on one side. Also called a hunter case back, it is made of solid material and works like a little round door when closed. To open it, you have to unlock it and gently lift it. This kind of system is also called an officer's case in reference to the first military wristwatches.

Indications and seals

The case back is usually the place to display important information about the watch. The most frequent data is the name of the brand, the model, the serial number, and the specific number of the watch if it happens to be from a limited edition. Technical data is also frequently mentioned: the water resistance with maximal depth, the presence of alternative materials, and other specifications. Watches made of precious material bear three authentication hallmarks for customs and legal purposes. The first denotes the kind of metal and its fineness, the second signifies the contractor, and the third, which came about due to the Vienna Convention, certifies the country of origin. For Switzerland, this last hallmark is a Saint Bernard head.

Straps and bracelets

The strap keeps the watch on the wrist. This mission might seem simple and nonessential but only if you have never had to deal with a bad strap. Straps can hurt your skin, and once this happens, you will quickly understand that straps are not a detail at all. The strap is a significant element of the watch as a whole because it has to be beautiful, well executed, and secure. Technically, a strap is mainly composed of two parts: each is linked to the case on one end, and they are secured together on the other thanks to a buckle that works much like a belt.

Of course, every watch is sold with one, and sometimes two, straps, like some diver's watches. Despite that, many watch aficionados and fashionistas have started to customize their watches by changing out their standard straps for custom-made examples. More than a trend, offering straps made on demand has become a peripheral industry in the watch business.

While "strap" and "bracelet" may feel like synonyms, rest assured they aren't. Straps are made of supple and usually organic material such as leather or a textile. Bracelets designate solid materials such as metal and ceramic.

Strap and bracelet materials

It is easy enough: You can use whatever you want as a material as long as the result is strong enough to resist the various physical constraints the strap or bracelet may be exposed to. A special craft, manufacturing leather straps is very demanding and may involve up to eighty steps. But the most important is that the products used to treat the material must be totally hypoallergenic.

Leather straps

Classy and natural, leather is a traditional material for straps. Accessible, long-lasting, and reliable, it has been employed in watchmaking since the rise of wristwatches. But the success of leather is also due to the fact that some animal skins are particularly rare and precious.

A leather strap is a refined, commonly handmade product whose manufacture involves many operations. Shape, thickness, color, finishing, threads, lining, holes, and feel are definitely important. Without entering into the specific details of fabrication, just note in particular that a strap is made of several layers providing a harmonious volume. Some brands like to have thick leather straps. Others prefer thinner ones. Leather straps, which can also be colored, are also finished with a clear seal to protect the leather.

Be aware that leather straps react to sweat and humidity. To be perfectly clear, after a certain point they can smell. To prevent this, try to not wear the strap on very hot days and avoid any humidity. Also consider that changing the strap is necessary roughly every two to four years.

Alligator skin straps
Today, the most common leather used for watch straps is alligator skin. Dedicated farms raise the animals before preparing their skins and sending them to strap producers. North America is among the biggest producers of alligator skins.

Each alligator skin is unique because of its design. The challenge for watch brands is to find and cut out the adequate parts from the main skin. This means that the scales must match in terms of shape and size to have a harmonious and coherent result.

Calfskin straps
Less prestigious than alligator leather, calfskin is appreciated by brands because of its comfort, look, colors, and, above all, strength. Indeed, naturally ocher brown, calfskin leather presents a smooth and homogeneous surface that can be also easily decorated and dyed.

Shell Cordovan strap
Prestigious in shoemaking, this horse leather, which comes from the rump of the animal, is also used for straps. The German brand Nomos introduced

it very early in its collections. Particularly resistant, Shell Cordovan leather may look less prestigious, but it is a very appropriate leather for watches because of its superior resistance to moisture and scratches.

Other leather straps
Many other leathers are also used: water buffalo, ostrich, shark, and shagreen—which is the polished skin of manta rays or dogfish.

Certifications and authentication of leather straps
Watchmakers have to source their leathers straps from authorized producers who must respect national and global legislation. The largest and most international of these is the CITES approved by eighty countries at a convention in Washington, DC on March 3, 1973. CITES stands for Convention on International Trade in Endangered Species of Wild Fauna and Flora. It is an agreement signed by the countries involved, whose goal is to control and monitor the exploitation of endangered animal skins. But, of course, this is not sufficient to ensure that all makers of leathers used worldwide for watches comply with this legislation. So, please, ask about the origin of the leather used on your watch strap when you buy it. And if the answer doesn't fit your expectation or the seller is unable to provide a certificate or official document, then avoid it.

Fig. 36: *Leather strap*

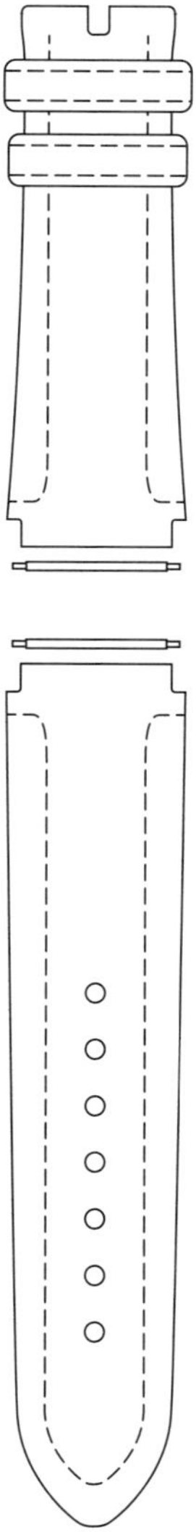

Fig. 37: *Metal bracelet (interconnected links)*

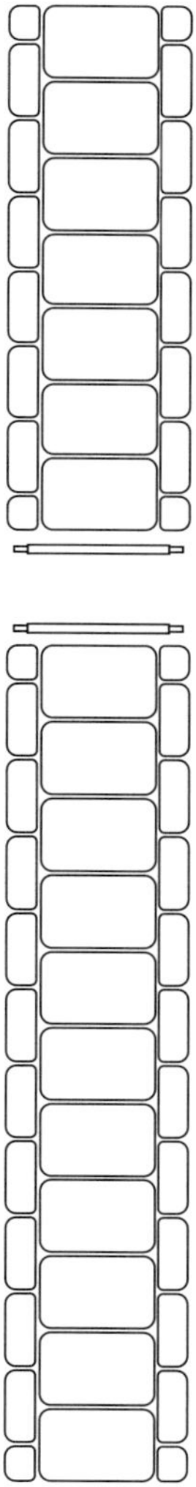

Metal and solid bracelets

Solid and long-lasting when well executed, metal bracelets constitute the second main category of watch bands. Metal bracelets, appearing later than leather straps in the history of watches, constitute an assembly of dozens of parts creating a kind of flexible puzzle. Some parts are visible while others are not; some also have mechanical purposes. Among them, you'll find pins, micro tubes, and screws.

The usual design for a metal bracelet (or any other solid material for that matter) comprises interconnected links. The quality and creativity of this "mosaic" gives the bracelet its style. Some have become symbolic of a collection, or even a brand, like the "Link" model by TAG Heuer.

Of course, the use of various metals, other strong materials, and adapted treatments allows a variety of colors. Sometimes, metal is used in the form of braided wires: the Milanese mesh bracelet is a good example.

Rubber strap

Another kind of strap is called rubber because the material used looks like the natural latex from the Pará tree. But even if the touch and feel are almost the same, the material actually is elastomer or elastic polymer. Silicone is one option.

The huge advantages of a quality rubber strap are evident in its comfort, reliability, and longevity. With the exception of voluntarily cutting it, there is almost no chance of damaging it. Even the dyes used have been improved in order to remain unchanged over time. Some rubber straps even resist the ink of a pen so that you can remove errant marks only by rubbing them.

Rubber was not a common material for luxury watch brands until Hublot introduced it in its straps at the beginning of the 1980s. Mainly used with sports watches, rubber straps are present in various categories of watches now.

The NATO strap

For a few years now the NATO strap has been a very fashionable addition even if it was first invented for military purposes. Indeed, the use of the acronym NATO signifies that this kind of strap has an official stock number with the North Atlantic Treaty Organization, which precisely defines its size and color. The institution created two references: one for the army and navy and another for the air force. But of course, textile straps are not exclusively reserved for the military anymore. Still used in professional environments, new NATO strap models were developed to fit around wet suits and space suits. Some make use of Velcro for fastening. But for the masses, NATO now rhymes with creativity, comfort, and the luxury of changing the look of one's watch rapidly. It has become a very easy and affordable way to personalize a watch.

Satin

Satin is a type of woven fabric. Historically, artisans worked with silk only, but now artificial materials like nylon and polyester are also employed. Watch brands often turn to satin for making feminine straps. Glossy and smooth, they are also appreciated because of the many colors available. But do be careful: the strap is easily stained if made of a natural material.

Buckles

The buckle is the indispensable system comprising three to twelve parts that joins the two sides of the strap on the underside of the wrist.
A buckle must be well conceived and executed to repel unexpected situations, stay locked all the time, and be ergonomic with respect to the wearer's skin. These are key elements, as a bad buckle has the power to entirely ruin the experience of wearing a watch; even the most luxurious watches are influenced by this. Finally, the buckle of a watch must reflect coherence with the overall design; it is fundamental to avoid having a plastic or steel buckle go along with a gold diamond-studded timepiece. Some carry the brand's logo, and very exceptional ones are even fully set with gems.

Fig. 38: *Pin buckle*

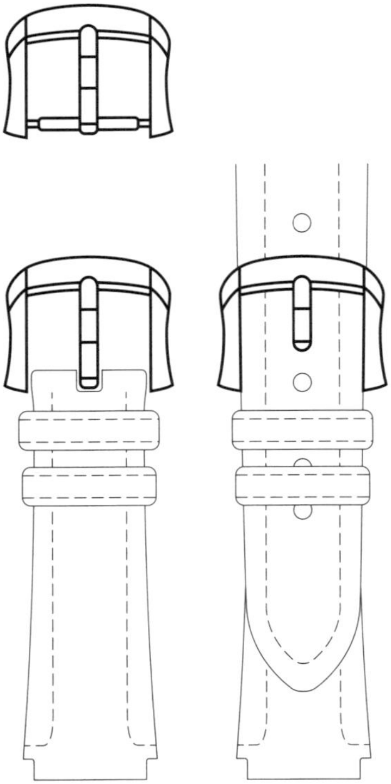

Pin buckle

The pin buckle is the simplest and oldest kind of buckle. It is composed of a pin encircled by a frame and attached to the strap by a bar.
To secure it, you just have to insert the other part of the strap into the frame, choose the right hole, introduce the pin, and pull it back. Think of closing a belt: this is the same. Of course, this kind of buckle only works with supple straps such as leather and rubber.

Its simplicity doesn't make the pin buckle trivial. On the contrary, considered very classic, the pin buckle is used on very rare timepieces as well as common, mass-produced ones.

Folding buckle

The folding buckle is a locking system that is firmly attached to both sides of the strap. It is usually made out of metal and contains a locking system. Thanks to this construction, the watch can't fall by accident while you are putting it on. It is also firmly secured on the wrist thanks to a security lock. The folding buckle preserves the strap from becoming worn out too early because no pin goes through it. A simple folding buckle is connected to each part of the strap and linked together by one hinge. When the wearer wants to close his strap, he pulls one side of the folding buckle onto the other one. The locking system keeps them together and the strap covers the metal strips underneath. Folding systems can work with two or more metal strips. Some don't lock because the blades are maintained on top of each other thanks to a spring system. These are known as butterfly clasps.

Fig. 39: *Folding buckle*

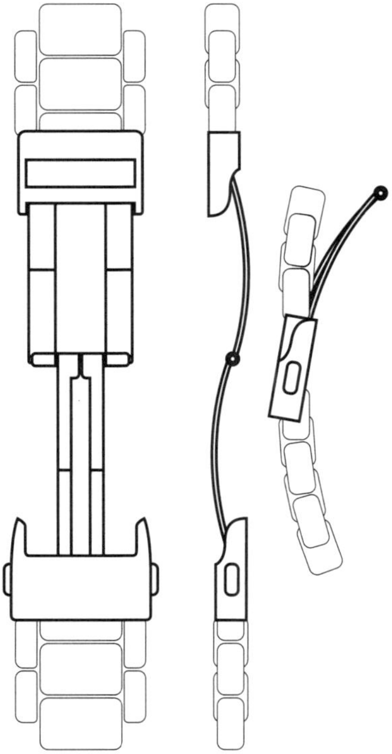

Complex folding buckles can also integrate extension systems. These are especially necessary for use with diver's watches. The watch must firmly stick to the wrist whether the wearer's arm is naked or covered by a diving suit.

No buckle

Chiefly used with digital or electronic watches, a strap can be a ribbon secured by a Velcro or other fixation system. These straps are not the norm and are employed mainly for technical and sports watches.

Materials

From wood to carbon fiber, many kinds of materials have been involved in the manufacture of a watch over the years. Some have proven to be the best for technical reasons. Others, like precious materials, are purposely utilized for prestige and looks. In recent years, and mostly because of the necessity to differentiate one's own products from those of competitors, brands started to work more than they used to with materials found in other industries such as auto racing, aviation, and space exploration.

Materials used in watchmaking must also respect strict specificities in order to be hypoallergenic, ensuring that they create no reactions on the wearer's skin. These are general standards respected by all established brands. But fake or low-quality watches can be more problematic because rough materials are sometimes not qualitative. Keep in mind that a skin reaction is possible; if your skin should react in some way, remove the watch and check with your doctor if you are not aware of being allergic to certain materials.

Sustainability has become a key word for brands and suppliers of the watch industry concerning materials. Almost all companies listed on stock exchanges include

this kind of best practice in their governance rules in order to maintain investor expectations. The main problem up to now concerns production operations and transformation of raw materials, in particular precious materials such as gemstones and gold. Various global agreements exist such as the CITES treaty from 1973 that aims to protect endangered species. Since 2003, the Kimberley Process fights against the trade of so-called conflict diamonds. In particular, it imposes control on diamonds' origins and whereabouts, but only in their rough state. Founded in 2005, the Responsible Jewellery Council creates standards and certification processes that concern the whole supply chain involved in jewelry, from the sources to the final product. In 2013, the RJC counted 470 members and had delivered 320 certifications.

All these efforts are naturally positive: they aim to reduce the importance of parallel trafficking and illegal activities. Nevertheless, the best test for you to be sure that the product you are willing to buy is sustainable is to directly ask the retailer and to get a clear, positive, and verifiable answer. If a watch reflects your personality, it should also embody your values.

Metal

Metal is used in any kind of watch. Some manufacturers use it only in the movement—even in the printed circuit boards—and some elements of the case, while others for almost all of the watch. Some watchmakers have also chosen to distinguish themselves by using precious metal to build movement parts. François-Paul Journe, for instance, manufactures the plates and bridges of his calibers in 18-karat red gold.

Some metals can also be coated with a layer of protection or a very thin coating of precious metal. The former concerns coatings such as PVD (physical vapor deposition) and DLC (diamond like carbon) that also allow color on the surfaces (exclusively black and dark gray for DLC). In the second case, generally a thin layer of gold is deposited on the component by means of electrolysis.

It is also possible to directly modify the structure (and color) of the metal by oxidation or heating. Aluminum, for instance, can be artificially oxidized. Called anodizing, the process creates a protective layer. This technique also offers a large range of colors. Heat treatment, used mainly with steel and titanium, transforms the metal to make it harder, also creating a (natural) protective layer, while coloring it. The famous "blued steel" used in watchmaking designates this technique frequently used on hands and certain parts of the movement.

Anatomy
Display
Indications
Functions and complications
Case
Straps and bracelets
Materials
Metal
Steel
Gold
Platinum
Silver
Titanium
Brass and nickel silver
Aluminum
Nonmetallics
Movement

As with titanium—a second application discovered and mainly used by De Bethune—this process implies rare skills as all the parts on the same watch must have the same hue regardless of size and shape.

Steel

Steel is the most common metal for watch elements such as the case and the bracelet. Steel is an alloy basically comprising iron mixed with a tiny amount of carbon.

Brands also frequently use stainless steel, which also bears the name “inox” in French-speaking countries as it comes from the French word “inoxydable” (stainless); its inventor was French metallurgist Pierre Berthier. Stainless steel sees at least 10.5 percent chromium (and at least 1.2 percent carbon) added to the alloy’s mass to make the substance more resistant to corrosion.

Today, more than 3,000 variations of steel exist on the market, but watchmakers only employ a few. Depending on the component they must build or their own experience, they have their preferences.

Gold

Malleable, resistant to corrosion, and above all magic, gold is utilized in watchmaking for parts ranging from bracelets to movement components. With time, gold can take on a patina that collectors in particular appreciate. But its softness has always been a worry because of its vulnerability to scratches and hits. Hublot developed an interesting solution: it invented Magic Gold, which mixes gold with ceramic material under very high pressure. Thus, components made of Magic Gold are virtually unscratchable because they take advantage of the extreme resistance brought to the alloy by the ceramic element.

Gold can be divided into different subtypes as described below. But for each of them, many nuances exist. They can be more or less intense and display different hues. Some brands have gone as far as developing their own types of gold by subtly playing with the compositions and proportions.

Yellow gold
Yellow gold has a bright, clear color. The alloy comprises 75 percent gold, 12.5 percent silver, and 12.5 percent copper.

Red gold
Red gold is very much in demand in luxury watches because of its warm tones. It is composed of 75 percent gold and 25 percent copper.

White gold
White gold showing no trace of yellow is made of 75 percent gold and 25 percent silver and palladium.

Platinum

Platinum is one of the rarest precious metals, and watchmakers use it for their most exclusive watches. Its structure makes it hard to machine and polish. It is used almost in a pure state (950/1000).

Silver

Silver is an option that costs less than gold or platinum. The only drawback is that its color changes with time. Some people might like this patina, but not everybody does.

Titanium

Light, very solid, and resistant to corrosion, titanium is in particular used for sports watches. Brands also like its hypoallergenic properties. Urwerk, in particular, has chosen to manufacture all its case backs in titanium because of its non-allergenic composition.

Brass and nickel silver

Brass and nickel silver—also known as German silver—are used to manufacture both stable and mobile components of the movement. Both alloys are made out of copper and zinc. German silver also contains nickel, which allows it to better resist oxidation.

Aluminum

Aluminum is appreciated for its lightness and resistance to corrosion. Nevertheless, it is a soft metal that needs to be transformed into an alloy for use; titanium, magnesium, and zirconium can be alloyed with aluminum in order to make it strong enough to endure the tough conditions a watch must sometimes go through.

Nonmetallics

Watchmakers also use many non-metallic materials. Some of these materials also have special technical properties in terms of lightness, resistance to shock, and magnetic fields—or they are used for aesthetic reasons.

Plastic

Plastic is important for watchmaking as a huge majority of entry-level watches make use of it. Plastic is a synthetic or partially synthetic material created from organic polymers that can be molded easily.

Nylon

Nylon is a type of thermoplastic invented in 1935 that withstands tension well and resists abrasion. It is mainly used to manufacture straps like NATOs.

Plexiglas®

Plexiglas® is the most famous and common synthetic crystal, and watchmakers use it to manufacture the crystals of some watches. Although you can still find it, nowadays sapphire crystal is the standard in crystals of high-end timepieces. Today, Plexiglas® can mainly be found on vintage, plastic, or fake watches.

Rubber
Rubber is a polymer used to create soft and flexible parts. Many straps are made out of a synthetic rubber such as silicone. The technology of rubber manufacturing has undergone major improvements, which means that almost any coloration is now possible and the material doesn't deteriorate over time.

Ceramic

Ceramic is a zircon powder that is heated and compressed in a mold in order to create a part. It can be also machine-tooled like any alloy. Ceramic material has been utilized for a few decades in watchmaking, and some brands have established their reputations using it, like Rado and even Chanel for its "J12 model." This material is appreciated for being unscratchable and for presenting intense and pure colors.

Sapphire crystal

Sapphire crystal is a type of industrial corundum that is obtained by working aluminum oxide at a very high temperature. Corundum exists in nature as a gemstone but can also be produced artificially with the same physical properties as a natural stone. Synthetic sapphire crystal is extremely resistant—it follows diamonds in hardness on the Mohs scale. Synthetic corundum can also be used inside the movement for rubies—the red color comes from adding chromium to the aluminum oxide of the normally colorless corundum in the manufacturing process. The pure version is used to manufacture crystal glasses. In this case, it is absolutely transparent and generally given an antireflective coating.

Carbon fiber

Carbon fiber is a strong, lightweight, black material. The fabrication process involves mixing carbon threads with an epoxy resin in a mold. The mold is then heated and the mixture put under pressure. The threads can be ordered in various constellations such as parallel, woven, and random before the molding operation takes place in order to create a special surface structure.

Silicon

The watchmaking industry has been using silicon to produce delicate parts of the movement such as escapement parts since the 2000s. The manufacturing process is very particular and implies use of a superior technology. This method can produce the high precision required for the most sensitive watchmaking parts for a minimum price.

Movement (Caliber)

It would be natural and intuitive to compare car engines to watch movements. But you just can't as they are not the same at all. Watch movements work continuously, day and night, for as long as they remain wound; this is not the case with car engines, which are started and stopped only when the vehicle is in use. The engineering challenges are completely different. A watch movement is thus a special invention with unique features. No other device can be directly compared to these self-sufficient mechanical constructions. Consequently, they must be analyzed and admired for themselves alone. And by the way: in terms of vocabulary, movement is the more common term, but caliber is a perfect synonym. Both mean the same thing today.

How it works

A watch movement transforms the energy contained in a battery or a tensioned metal spring into regular impulses. That is why every watch "beats" like a human heart does. These impulses are transmitted to the display system, which can be hands, disks, or a screen. As such, a movement is a machine that transforms the flow of time into beats that can be counted and displayed.

A movement is a small machine consisting of five major units. The first one is the power unit. Power is stored in a spring located within a container called a barrel in a mechanical movement or in a battery in an electric movement. The second is the transmission system comprising gears and called the gear train. The third one is called the escapement. The escapement is a sensitive mechanism that transforms still constant energy into regular impulses. It creates the tick-tock sounds you hear in mechanical watches. But this escapement must work in a perfectly regular manner in order for the watch to be precise. This is why it is connected to the balance wheel with its hairspring, which is the fourth unit. The fifth and last unit is the display.

Energy

On a watch, energy is provided via a battery or a mechanical spring, but there is a major difference between them. The battery always delivers the same quantity of energy. On the contrary, the concentric metal spring can't even pretend to have the same sort of regularity. Indeed, it releases a lot of strength at its maximal tension. Then the flow is somewhat stable before ending rather abruptly. Engineers and watchmakers must take this behavior into account at the time they conceive and work on mechanical calibers. Even if the regulation system can absorb some variations, all the watchmakers' talent is expressed by achieving the most stable energy flow from the beginning until (almost) the end.

Sometimes on mechanical watches, the level of energy left in the mainspring is displayed on the dial in the form of a power reserve indicator. But this is not a given, and sometimes the only way you know you are out of power is when your watch stops.

Transmission

A watch movement uses its energy to animate the time display so that it shows the correct time. In electronic (quartz) watches, the current goes through microprocessors and printed circuit boards. In mechanical movements, a system of gears usually comprising three to four brass gear wheels fulfills this mission by connecting the spring barrel (energy source) with the regulator, which

Anatomy
Display
Indications
Functions and complications
Case
Straps and bracelets
Materials
Movement
How it works
Energy
Transmission
Escapement and regulation
Frequency
Indication of time
Limits and weaknesses
Electricity
Mechanical

portions the energy correctly. Each wheel's axis is a staff made out of steel formed as a pinion, which contains a set of teeth surrounding the staff to drive the neighboring wheel in order to obtain the gearing. Each staff is capped at both extremities by a small pivot revolving within a cavity made of synthetic ruby (this synthetic ruby part is called a jewel in technical horological parlance). The rotation speed of each wheel corresponds de facto to an indication of time: the hours, the minutes, and the seconds. Thus, some watches, like those that show the time in regulator style, see a hand directly attached to each of these axes.

Escapement and regulation

The escapement, which transforms the energy into beats, is part of the regulator ensemble to control the uniformity of the energy portioning. The entire regulator can be regarded as the control center of the movement as it has a major influence on the watch's precision. Metaphorically, it is often called the "heart" of the movement, especially in mechanical watches, thanks to the pulsing motion of the balance spring. But the comparison stops there. The beats of an organic heart vary based on physical activity. In a watch, it is just the opposite: variation is the thing that watchmakers strive to ban.

Frequency

Frequency expresses how many times the regulating system makes the same cycle in the space of one second. Expressed in the internationally recognized unit Hertz, it is determined by the sum of semi-oscillations per hour. Watchmakers consider an oscillation (or "beat"), as the sum of two distinctive actions, the first going in one direction, the second in the opposite—like a teeterboard. Each of these is called a semi-oscillation, one oscillation being the sum of the back-and-forth motion.

Have you ever listened to your (mechanical) watch by putting it to your ear? Then you must have heard a little "tick-tock," right? This is the sound of minuscule parts going back and forth to divide the energy.

Frequency is a major element for precision because the higher it is, the more a period of time will be divided into short impulses. And the more impulses you have, the more precise a measure you get.

Additionally, every mobile mechanism is more stable if it works at a higher speed. To put this into perspective, consider how it might have been at the beginning of the twentieth century if chronometers had only been able to measure in seconds and not hundredths of seconds to time and establish rankings of professional skiers in competition? That said, using high frequencies implies more energy consumption, the need for better lubrication to reduce to friction between parts, and mastering the faster wear and tear of components.

Around 2010, brands such as TAG Heuer and De Bethune presented completely new escapement systems using magnetism. De Bethune went even further by introducing an experimental system to replace oscillations with magnetically controlled vibrations based on the physical phenomenon of resonance. These systems, which are mostly experimental, have the advantage of eliminating physical contact between parts, which represents a major step forward since friction limits possible speed.

Indication of time

Time can be shown in many ways as we discussed in the "Display" section. In electrical and electronic watches, micro motors controlled by the microprocessor move the hands. In mechanical timepieces, they are powered by a system of cogwheels (the gear train). When disks or other kinds of indicators are used, alternative and more complex systems assume the task of managing them.

Limits and weaknesses

From the time that blacksmiths and locksmiths built the first timing mechanism until engineers created movements that can resist accelerations up to 500 Gs, many improvements have been conceived concerning the quality and resistance of watches. Nevertheless a movement—whether electronic or mechanical—remains a sensitive and delicate object.

Anatomy

Mechanical movements are proportionally more fragile than the electronic variety for the simple reason that they contain more mobile elements. And the more there is—as with complicated watches—the more the probability of failure increases. Over time, of course, these parts naturally wear out and need to be fixed or replaced.

That said, a few key factors might alter the ability of a movement to function properly. Shocks are problematic. Even if great advances have been made, no watch can resist repeated or major shocks. And even if the movement holds up, other elements such as the hands, the dial, and the crystal can break, be damaged, and change position. Variations in temperature can influence a movement's rate, making it run faster or slower—which impacts accuracy. Dust and impurities also perturb a smooth rate. But humidity is certainly the worst factor: nasty, invisible, hard to detect, it makes metal rust, the crystal show condensation, surfaces oxidize, damages lacquer on dials, and corrodes the connection to the battery.

Electricity

As soon as they could—which was at the beginning of the nineteenth century—watchmakers worked on using electricity to power watches. They wanted to take advantage of this new, more regular, and controllable energy.

As we've discussed, to display time the challenge is to transform a flow of energy into impulses. The path to doing this in electric watches started with the first electric clocks. The goal in clocks was to control the movement of the pendulum. To do so, a piece of soft iron—a very pure iron with particular magnetic properties—was positioned at the end of the pendulum. Underneath, an electromagnet controls its motion by providing regular electrical pulses. Later it appeared as though the shape of a tuning fork would be optimal; prongs vibrate regularly when electricity flows through them. The vibrations could then be used to directly regulate a gear train or to give the right pulses to a microprocessor. The tuning fork shape turned out to be the right one as it remains the norm today.

Quartz

Quartz is a hard mineral composed of silica and oxygen. It also has a special place in the world of watches since it has been used to run electronic calibers for more than fifty years: quartz is not only a nice, shiny stone to decorate the living room table. It has the very particular physical property of being piezoelectric. This unusual word means that the crystal extends or retracts when an electrical current goes through it. And the phenomenon goes in the other direction, too: push, extend, or twist a piece of quartz and one side becomes positively charged while the other becomes negatively charged. This motion can be really fast—so fast that you can have thousands in a single second to create the desired vibration. At this point you have to remember that, to measure time, you must transform linear energy into impulses. And again, the more you can divide the period of time, the more precise it becomes. Quartz is then a great ally of precision because its vibrations are stable and uniform once electricity passes through it. This is why engineers and watchmakers worked hard starting at the beginning of the twentieth century to make quartz a part of watch movements—and they were successful.

So how do you convert the superhigh 32,768 Hz vibrations of a tiny piece of crystal into a time display? By using a frequency divider circuit. This system commutes the vibrations into impulses sent to a tiny electric motor, which powers the hands. On digital displays, the impulses directly activate the segments of the screen.

Analog quartz

Focusing on the concept of tradition mixed with computing, watch brands have developed very interesting and creative products combining analog time display and screens for additional information. Measuring gravity, pressure, temperature, depth, distance, and even displaying compass directions, this world seems limitless as appropriate sensors have been developed to fit into a watch and provide accurate data.

Casio might be one good example of a brand inventing creative functionalities using quartz technology. Its "G-Shock" collection contains analog models with a wide choice of unusual functions displayed on a screen. Launched in 1983, the collection offers three core elements: resistance to shock from a fall of 10 meters, water resistance to 10 bar, and a 10-year battery life.

Mecaquartz

Some brands have cleverly decided to merge quartz technology with traditional fine mechanics to take advantage of the precision of the first and the prestige of the second. The concept is simple: energy is provided mechanically thanks to a spring barrel connecting to an automatic winding system. A generator then converts the energy into electric power and transmits it to a quartz regulator. As in any quartz movement, the regulator activates micro motors that drive the hands. But the power can also animate complications built in a traditional mechanical way. Such calibers are bigger than usual quartz movements because of the mechanical modules.

The Japanese brand Seiko has created a whole collection named Kinetic using this kind of functional architecture.

Tactile quartz

With tactile quartz models the case and/or the crystal can be used to control functions. Omega and Rado have had watches in their collections with contact-sensitive cases. Their Swatch Group sister brand Tissot, in particular, took advantage of quartz and microelectronic engineering to create the "T-Touch" in 1999. Its crystal is actually sensitive to tactile pressure. Since its launch, the "T-Touch" has come out in many versions dedicated to various environments and sports such as diving and alpine activities.

Rarities

Sometimes engineers appreciate pushing quartz and microelectronic science to its limits in order to invent extremely high-tech watches. Pretty rare and expensive, these models have been designed to offer outstanding functions. Here are three examples of such rarities.

Breitling has a long history in aeronautics, and that is certainly why in 1988 the brand launched a watch containing a miniaturized emergency radio transmitter able to broadcast on civil aviation's distress frequency of 121.5 MHz. Appropriately named Emergency, the second version of this watch launched in 2013 emits a secondary signal at 406.04 MHz monitored by satellites. In 1990 Junghans's engineers implemented a radio-controlled system in a watch in order to catch the signal from the Frankfurt's atomic clock. Called the "Mega 1," its antenna is integrated into the strap.

And what if you desired the sun to provide you with the time? Citizen has indirectly managed this with its Eco-Drive system, first presented in 1995. Thanks to a special dial able to absorb light and transform it into electrical current, Eco-Drive quartz watches no longer need batteries. Any source of light, natural or not, is enough to fill the bill.

Mechanical

In a mechanical movement, it is all about wheels, gears, springs, jewel bearings, and oils working together with only the physical laws of mechanics to govern the interactions between all the components. This rare complexity mastered by the human brain and hands creates all the appeal, the mystery, and the prestige of mechanical watches.

Mechanical watches continue to fascinate for many reasons, starting with the amazingly small size of their elements. Indeed, the scale of measure here is a micron, or one-thousandth of a millimeter (0.001 mm). To understand a little better what that means, consider that one of your hairs measures between 0.06 to 0.07 millimeters. Until the nineteenth century and the improvement of machining tools, building a watch was a craft in which only specialists were able to design the movement, manufacture the components, assemble them, and make it all work. Such individuals still live among us today even if they are very rare. In fact, each of their tasks has become a proper field in itself with a great level of specialization. The complexity of movements has also increased significantly, making it sometimes all but impossible to apply the traditional ways of conceiving and creating watches using pencils, paper, and blueprints. Calibers are now designed by movement designers and engineers using computers similar to those found in automotive engineering. Micro mechanics are in charge of programming the computer numerical control (CNC) machining tools, which carry out the actual production of parts. Then watchmakers finish, assemble, test, and adjust the movements before fitting them into cases, adding the dial and hands, and closing them up. Because of this long, complex process, many tests are done to check the quality at different stages. Dedicated craftspeople can also execute the different tasks separately. For instance, the assembly of a movement can be divided into subgroups of parts with specialists in charge of assembling them. The experienced watchmaker focuses only on the most sensitive operations.

Art and decorative craftsmanship also find their way into movements. Mirror polishing, beveling, engraving, and even gem setting are the usual techniques applied in fine watchmaking to embellish the mechanics by creating visual contrasts between the various parts and their surfaces.

Even if it is absolutely true that the principles of their design haven't significantly changed since the nineteenth century, watchmakers and engineers always work on improving the reliability and precision of movements. In this quest, they have perpetually integrated innovations at every level, whether with materials or engineering processes. Nowadays, experimental techniques have emerged that, for instance, allow moving parts to be maintained in an almost floating manner thanks to micro magnets creating a small magnetic field. This helps reduce friction between classic metal parts. Silicon has made more and more inroads in the fabrication of crucially sensitive and thin elements like the hairspring, the pallet lever, and the escape wheel. Some parts are also made out of carbon fiber or ceramics—for instance with ball bearings—and other parts are coated with a layer of artificially grown diamond for protection. Very rarely, these new materials allow for new mechanical systems. The "Constant Escapement" from Girard-Perregaux introduced the physical principle of buckling in watchmaking thanks to the use of silicon technology.

Structure of a mechanical movement

A mechanical movement can be imagined like a sandwich with a lower part and an upper part dedicated to securing all the moving parts (like wheels) between them.

Base plate

The base plate is the stable base of the movement. It usually looks like a flat coin with plenty of cavities and holes in it to allow space for moving components and screws.

Moving elements

The moving parts are mainly located in the heart of the movement. At times, some can be placed outside the base plate to highlight them.

Anatomy
Display
Indications
Functions and complications
Case
Straps and bracelets
Materials
Movement
How it works
Electricity
Mechanical
Structure of a mechanical movement
Functioning
Basic vs. complicated movement
Integrated vs. modular movement
Skeletonized and openworked movement
Finishing and movement decoration

This is the case, for instance, with the "Legacy Machines" collection by MB&F in which the balance wheel and hairspring couple is placed above the dial.

Bridges

Bridges are sandwiched on top of the moving parts. They can also have holes for screws to go through that secure them to the base plate. The shapes of bridges and their configurations are like pictures on ID cards. Each says a lot about the movement, and connoisseurs are able to recognize certain movements and watches by only having a look at their design and specific elements. You may be able to do that one day!

Functioning

A mechanical movement is basically a system composed of interconnected cogwheels. The forms of their teeth as well as their diameters and sizes generally differ based on the roles they play. To fit into a case, this set of components assembled much like a puzzle is conceived to occupy a minimum amount of spatial volume. That is why a caliber can look so complex and confusing. Without this constraint, a mechanical movement could be constructed in a line, for instance, with each element put end to end next to each other. At one end, you would find the drum barrel (alias the energy store); in the middle a series of a minimum of three cogwheels forming the gear train (for energy transmission); and at the other end the escapement and balance for portioning the energy into segments.

Power

As a child, did you ever play with toy cars that had to be wound using a key to run around the floor until their energy was used up—or have you at least seen these in a television ad? Basically, a mechanical watch works the same way, except that the energy lasts much longer because of the regulator subassembly that portions it. (A toy car does not have this regulator, which is why the energy runs out so quickly).

In a mechanical watch, energy is stored in the mainspring, which is located inside the spring barrel, a small round container that revolves around an axis called the barrel arbor; it drives the gear train. Made out of a special metal alloy, the mainspring is created with a very specific manufacturing process. Its mission is to unwind, releasing tension and thus energy as regularly as possible no matter the conditions and the level of tension. At one end, the spring is fixed to the center of the barrel arbor. The barrel arbor is the part that rolls the spring up around itself, thereby creating the tension. It is activated by the barrel's ratchet wheel, which is a cogwheel placed outside the barrel and directly linked with the manual or automatic winding system.

At the other end, the mainspring is fixed on the side of the barrel drum that powers the gear train thanks to an external toothing. The mainspring can end with a slip spring that glides along the barrel to avoid it breaking from excess winding. That system is often found in older watches; in newer timepieces the mainspring is usually secured to the barrel.

— Power reserve

In a mechanical movement, the available power reserve (the number of hours the watch is supplied with energy) depends on various factors. The length of the mainspring has a significant impact. And in certain watches, the mainspring can measure more than one meter! The "8 Days movement" from Eberhard & Co. has one of these, for example: its mainspring measures a total length of 1.55 meters. Watchmakers can also add several spring barrels, a feat that can be compared to putting several gas tanks in a car. The "MP-05 LaFerrari" by Hublot, for example, houses eleven barrels for a total power reserve of fifty days—more than a month and a half! In comparison, a regular automatic caliber usually has a power reserve of about 42 hours. The winding system has a direct influence on the conception of the caliber (you can read about the differences between systems below). Remember that a longer power reserve on a manual winding watch is a positive thing: If the owner must wind a watch less frequently, it saves the mechanism from excessive wear and tear. The benchmark set for some years now is eight days of power reserve. Why such a length? First, it makes sense because some aficionados own more than one watch and generally change them on the weekend. As they don't want to have to set a watch each time they pick it up, they like to have a power reserve that lasts for at least one week.

Last but not least: Keep in mind that the duration of the power reserve provided by the brand usually corresponds to the period of time during in which the torque is strong enough to ensure the precision of the watch, but that it will most likely run longer.

— Manual winding

The manual winding system was the first one invented, and it is the simplest version: Energy enters the movement from an external source. In old watches, owners needed a key or another tool to wind the movement. That is still the case with many wall clocks, for instance. In 1820, an English watchmaker named John Arnold invented a little external element that allows the wearer to wind the barrel: the crown. The gold standard today, the crown is generally positioned on the same side of the watch as the key spring barrel elements. Note that you can wind every mechanical movement—even automatic examples—using the crown.

— Automatic winding

Automatic movements are as clever as they are practical. It's true, turning your crown regularly to wind the mainspring can be boring and take up your time—and it goes without saying that if you forget to do it, your watch will stop working without you noticing it. Comfort and a need for a certain security regarding energy reserves inside the caliber motivated watchmakers to invent self-winding systems. The first watches including a self-winding system were invented at the end of the eighteenth century. But today, automatic watches outnumber manually wound ones and with good reason.

From the beginning the idea was to use the motion of the owner's body, and in particular his or her wrist, to generate energy. To do so, watchmakers added a mobile element to the back of the movement that oscillates back and forth. This part, called the rotor or the oscillating mass, has the approximate shape of half a disk and revolves around an axis. The rotor is directly connected to the spring barrel via a gear. Some systems wind only in one direction of oscillation, others in both; this is a choice that the movement designer makes. The result is the same, except that it takes a little bit more time to fully wind a movement with a unidirectional system. On certain movements the rotor makes some noise when it moves, which can be considered a kind of signature element. And if you pay close attention, you can usually also physically feel the motion of the rotor when you wear the watch.

Note that most of the time the axis for the rotor is situated in the center of the movement—but not always. Some models, and only some, are equipped with micro rotors. Micro rotors are smaller examples that make their revolutions not above the movement, but within it. These are chiefly created to reduce the height of a movement and are much more challenging to manufacture and adjust.

The regulation: escapement and oscillator

The goal of the mechanism usually summed up as "the escapement" is to regulate the flow of energy coming from the barrel through the gear train and to split it into regular impulses to count time. You can see this action played out on the dial. Indeed, if you have a long second hand on your mechanical watch, just note that each almost imperceptible step it takes forward is the result of one impulse. The regulating system is a set of parts very accurately made and adjusted; its precision is passed on to the act of time measurement.

Strictly speaking, the regulator is divided into two major subassemblies: the oscillator (balance) and the escapement. The oscillator acts in combination with a large, light wheel called the balance wheel and an ultrafine hairspring that is rolled into coils and located in the middle. These two components are the most delicate and sensitive in any mechanical watch. Their manufacture is the biggest secret brands hold on to, and professionals able to work on them are rare, gifted, and very much in demand in the industry. One end of the hairspring is secured to the balance staff, while its other extremity is secured to an external and immovable attachment point. Its role is then to oscillate, meaning turning in one direction before coming back in the other in order to create impulses. The escapement subassembly comprises the escape wheel, pallet lever,

Fig. 40: *Escapement*

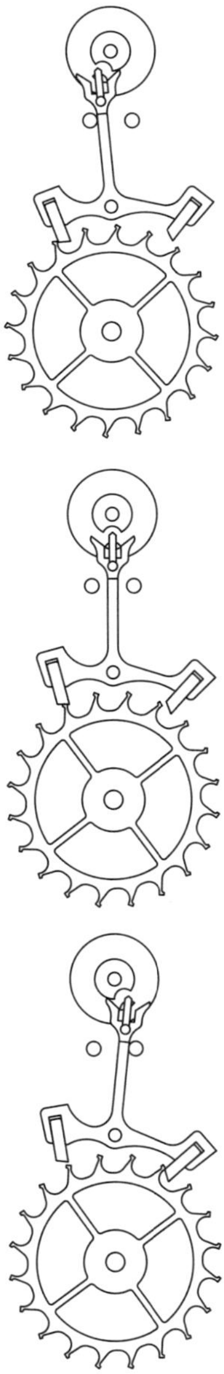

and pallets. It portions the energy and sends it to the oscillator. Working together, the oscillator and the escapement transform the linear flux of energy into pulsations that give rhythm to the entire movement. The speed of these oscillations provides the frequency of the caliber. And the higher it is, the more precise it becomes.

Jewels and shock absorber

The various wheels and components of a mechanical movement must turn as easily as they can; any friction represents a loss of energy and precision. For this reason, watchmakers use synthetic rubies called jewel bearings to maintain the ends of the axes with as little friction as possible. These jewels are minuscule carved disks with or without a tiny hole in the middle. Historically, they were made out of natural ruby; now they are the purer synthetic ruby. Thanks to their man-made manufacture, colors can also vary from the original red. Usually, the number of jewels a movement contains is engraved on the movement.

The rubies maintaining the most sensitive axes—like the balance wheel, for instance—are usually doubled with a shock absorption system. This component mainly has the shape of a circular micro blade spring positioned within a fixed part where the jewel is located.

Basic vs. complicated movement

From a relatively basic mechanism a movement can grow to become a very complicated construction. This depends directly on the number of functions it contains. A standard movement has approximately 150 parts. This number can go up to more that 1,000 on really complex calibers. As each component is a challenge to manufacture, assemble, and adjust, it is common to refer to calibers as basic or complicated movements.

Integrated vs. modular movement

Conceiving a complicated movement requires more effort and means than a basic one. For this reason, other solutions were invented to reach a level of complication in an easier and more practical way. Today, watchmakers have the choice between two different approaches to create such movements.

The first one—the most logical—is to develop the movement with the complication built directly into it. Called an "integrated" complication, this is a movement that sees the complication(s) as an intrinsic part of the caliber. This kind of construction is more valued by horological purists. They like that the caliber can't be modified and that its original form comprises the additional function(s). The movement is holistically one.

The second way is to create a complication module and add it to a base movement. The base movement is the source of energy and drives the complication module. This solution is still very current as conceiving a whole, complicated movement is extremely expensive (starting at about $2 million) and requires many years (in general up to five before the caliber is totally reliable). For these reasons, and because developing a module is also less expensive, many brands use this solution.

Skeletonized and openworked movement

Some watches have no dial—and this is not a sign of laziness. On the contrary, watches whose calibers are visible require more work and attention. Called skeletonized or openworked watches, depending on how they have been produced, their goal is to magnify the mechanics, expose the heart of the caliber, and also prove to the customer that the brand masters real know-how.

Watches without dials are not recent. André-Charles Caron (1697–1775), royal watchmaker to the French King Louis XV, is considered to be the pioneer of this style. Over time, clocks, pocket watches, and wristwatches have all been realized in skeletonized versions. But how do watchmakers do it?

Traditionally, the goal was to put the structure of the movement on show, thereby underscoring its complexity. In any case, a central goal of the operation is to make the components beautiful to look at. Therefore they can be polished or brushed, their edges filed and beveled, and sometimes engraved to make each part very perfect. Sometimes they also receive a coating to color them. Also keep in mind that an important goal of skeletonization and openworking is to allow light to enter the movement as much as possible. To that end, watchmakers sculpt the components before finishing them. This process sees the watchmaker manually removing a maximum of material from each part using a special saw and file. An additional goal is to express some creativity by giving a new and elegant shape to the parts using engraving and guilloché techniques. In all of this, one important objective was also to demonstrate superior ability.

Today, movements can be directly created in a "skeletonized" shape. However, these cannot literally be considered true skeletonized movements because the parts are directly produced in the final shape. The watchmakers are spared having to cut and reshape each element and can't express their style through these operations.

Consequently, it would be adequate to call them "openworked" movements. And even if they are different from the traditional kind, they offer a lot of novelty. Actually, they can represent an image or the numerals can be integrated directly into the movement like the Cartier "Santos 100 Skeleton" offers.

Finishing and movement decoration

Fine watchmaking will forever be a mix of technology and beauty. This is why movements are sometimes decorated to be as beautiful as they are efficient. Of course, decorating a movement involves additional costs and extended production schedules. That is why different ranges of decoration exist. From a strict minimum to polishing parts the size of half a matchstick for a whole week, a lot of variation exists.

Watchmakers originally finished their movements in order to erase traces left by machines and tools during the production process. That is still true today. But finishes have also been developed as an art form in themselves. In addition, watchmakers realized they could create visual effects by alternating different kinds of finishing on parts. By giving a pattern to surfaces or making others shine like mirrors, they have found ways to transform metal mechanisms. And even if most components decorated like this are partially, or sometimes totally, hidden, it remains an obligatory hallmark of high-end watchmaking.

Getting introduced to the full range of horological finishes takes time because it implies knowledge and experience. And before I forget: finishing and decoration also make for fantastic discussion topics cherished by watch aficionados.

Fig. 41: *Beveling*

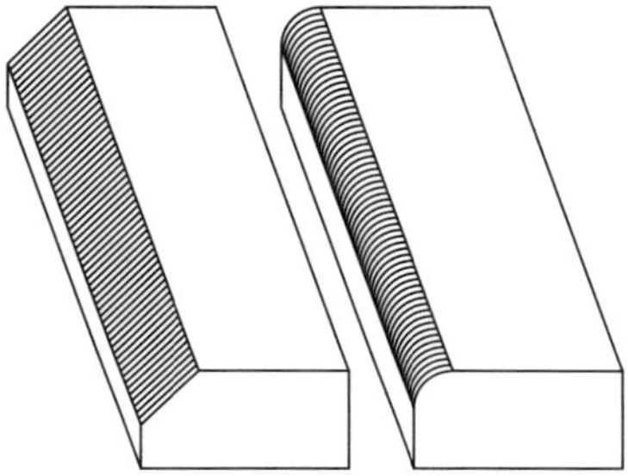

Beveling

Beveling a part consists of cutting the edge between its surface and its side to a determined angle—ideally 45 degrees. This has two advantages. First, it gives the component a smoother look. Indeed, an untreated part appears rough and lifeless. Beveling makes it less harsh. Secondly, the bevel is at least shiny and sometimes also specially polished. Consequently, it participates in creating visual play in the movement.

Beveling is a handcraft but can also be executed by machines. The difference shows up in how purely it is done, how intensively the angles reflect the light, and if the angles are flat or domed. Sharp internal angles must have a clear little caesura in the middle only achieved by human hands.

Fig. 42: *Circular graining*

Circular graining

Circular graining (also known as "perlage") consists of decorating a metal surface with small round patterns between 2 and 5 millimeters in diameter. These rounds are positioned regularly and partially juxtaposed. Generally, they form concentric rows starting from the center of the movement. Machines can produce them, but usually an artisan places the circles on the components using a hand-operated tool with an abrasive, round, rotating peg. The operator pulls the peg down onto the surface of the part to create a circle before moving the part slightly and make the next one, etc.

Fig. 43: *Geneva stripes*

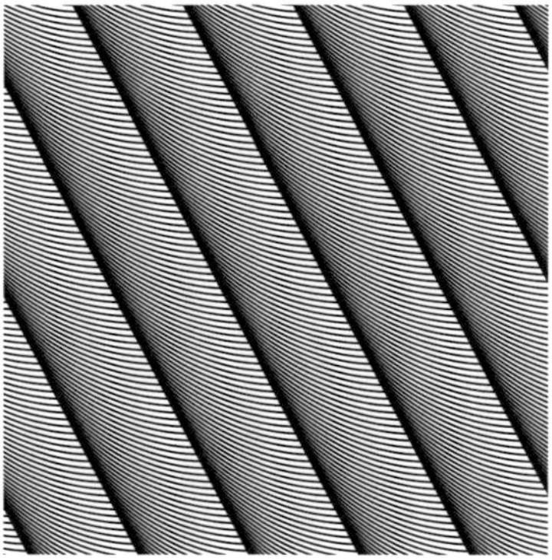

Geneva stripes

Geneva stripes—very often called "côtes de Genève" as in French or Geneva waves—are series of slightly concave stripes made on the surface of the component. Most of the time, they are linear and parallel. But certain brands use arabesque patterns; other brands use round ones. They can become a distinctive feature such as with German watches from the Dresden and Glashütte areas, where this decoration is used in particular to adorn the spring barrel lid, the plates, and bridges of the movement. In that region, this decoration is called Glashütte ribbing.

Geneva stripes can also be applied in various widths. Sometimes thin ones alternate with wider ones following an established pattern. In addition to its width and depth, each stripe shows regular and diagonal micro rays due to the tooling. Some experts say that Geneva stripes have the ability to retain particles of dust to prevent them from going elsewhere in the movement.

Satin finish/brushed finish

Satin finish describes surfaces that have been finely and conscientiously rubbed with an abrasive paper or an equivalent tool. Note that satin finishing is also used on cases and dials.

— Straight polish

The most common version of satin finish is a straight polish. It consists of parallel micro rays that cover the entire surface.

Fig. 44: *Satin finish*

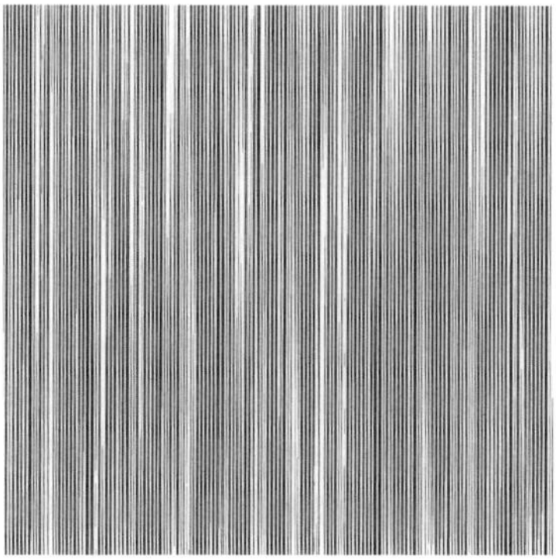

Fig. 45: *Sunray*

Fig. 46: *Snailing*

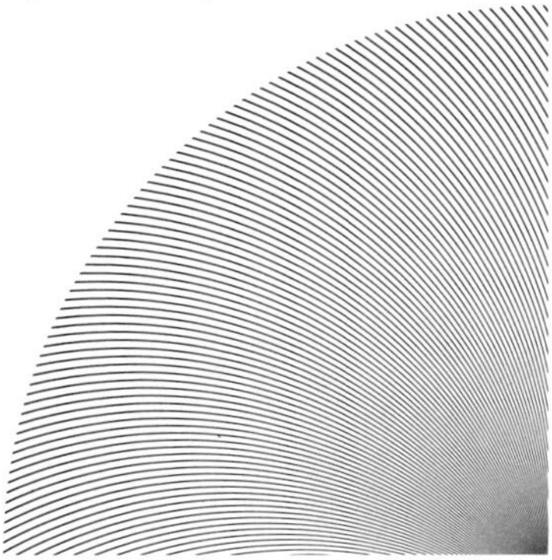

— **Sunray**
Sunray satin finish is basically used to decorate round parts. In this case, the micro rays start in the center to extend to the edge like a ray of light bursting from the sun.

— **Snailing**
This word describes a type of decoration similar to sunray satin finish. The difference is that the rays, or lines, are shaped in a spiral extending from the center.

Grainage
Grainage, or sanded, finishing is used to create a uniform matte surface that evokes a crystalline structure.

Polishing (mirror finish)
The polishing of parts carries major importance in high-end watches. The goal is to make a surface shine like a mirror. Many tools, hours of work, and experience are necessary to succeed in this task.

Engraved
Movements usually display some information engraved on them. Very often, you can read the name of the brand and some technical data such as the reference number of the movement and the number of jewels. Normally, lasers are used to make these engravings. These mustn't be confused with artistic engraving, which is purely decorative and much more difficult and time-consuming to do.

Conclusion

You have just discovered how fascinating watches can be if you just take the time to get to know them better. Congratulations and thank you for your curiosity. From now on, it can only get bigger. This was the goal of *The Magic of Watches*, which was conceived as the beginning of an initiatory journey into a special art and into total freedom.

Watchmaking and wristwatches express such a variety of styles and tastes, it was not possible—nor even desired—to illustrate the whole picture. In addition, personal subjectivity often matters more than the facts themselves. That said, the knowledge you have acquired will certainly help you often, whether you are a keen amateur or have since become much more addicted to these little beating objects. In any case, continue to nurture this idyll because it is worth it.

It is often said that everything has been invented in watchmaking. This statement is somewhat true. Yet watches remain current because brands and watchmakers draw on this heritage to constantly reinvent themselves. They also work hard on cultivating a taste for surprise and wonder, which merits close and regular monitoring of their news. As a springboard, *The Magic of Watches* was designed for that purpose: to capture and understand, at any time, the world of watches, based on solid foundations.

Thus, *The Magic of Watches* forms a set of keys, born of a desire to share, of a passion for this special art, of beautiful friendships and the goodwill of everyone who was involved in its creation. As the author, I gathered the knowledge and skills of those experts and specialists to create a book designed to be useful and meaningful.

My wish is now that you look attentively and with enthusiasm at the creations of watchmakers. As with any little treasure, every watch deserves time in order to discover its value and beauty.

Sponsors

Fondation de Famille Sandoz
Fondation de la Haute Horlogerie
André Gigon
Eric Giroud
Enrique Grebler
Pascal Raffy
Jean-Marc et Catherine Wiederrecht

Acknowledgments

"The Magic of Watches" has received various and precious support, starting with my family, relatives and friends. They made possible to create this guide of a particular kind. The sponsors who financed its realization deserve a special mention as they showed trust towards the project and its author. Different experts also graciously helped by revising the manuscript. Finally, there is all those who transmitted an incredible energy by their encouragements and enthusiasm. This momentum has been the real source of power of "The Magic of Watches," consequently I thank all them warmly.

In addition to the sponsors, the birth of "The Magic of Watches" was made particularly possible thanks to the involvement of:

Sébastien Chaulmontet
Elizabeth Doerr
Gianfranco Ritschel

Daniel André
Aurel Bacs
Josée Bélanger
Patrick Berdoz
Anne Bieler
Florence Bourqui
Marc-André Deschoux
Christian Ferrier
Carlo Lamprecht
Philippe Lebet
Fabienne Lupo
Anders Modig
Fabrice Mugnier
Marc-Christian Perronnet
Anthony Randall
Lucien Vouillamoz

Ahmed Seddiqi and Sons
AIM Services SA / Tell Softwares
Greubel Forsey SA
Haldimann Horology
Laurent Ferrier SA
Virtualideas Sàrl

Credits

The Magic of Watches
A Smart Introduction to Fine Watchmaking

Consulting and Editing: Elizabeth Doerr
Consulting: Sébastien Chaulmontet
Consulting: Gianfranco Ritschel

English copyediting and proofreading: Ashley Benning, USA

Graphic design: Bogsch & Bacco, Switzerland

Illustrations *Icons*: Virnard Magpantay, USA

Poster: Universal History Archive/UIG via Getty Images

Printed by Graphius, Belgium

ISBN 978-2-940506-12-5

Index

Extensive table of contents

Alphabetical order

Notes